HOW TO MAKE YOUR BAR PA KICK BUTT

JERRY R. DEPEW

FIRST · EDITION

PREFACE

Dear Reader,

I started writing this book back in 2015-2016. It didn't take me very long to write the first three quarters of this book because I had been thinking about this topic for a long time. I had a head full of information that I wanted to share and get out on paper. I was super excited about writing a book on how to make a sound system kick butt and as time went on, things in life happen. I would end up not liking a certain part I wrote or I would learn more information and make changes. I even re-wrote & deleted chapters entirely because I wasn't satisfied. As life would have it, little fires would crop up and the next thing I know, the book is put on the back burner while I take care of something else. This cycle repeated itself several times.

I don't know if any of you have ever written a book before but it can be a daunting task. Extremely time consuming, repetitive, looking for errors, researching some more, life's little fires popping up, raising a family and on and on. When it became evident to me that "analog" was being replaced with digital equipment and powered speakers, I nearly quit trying to finish this book. Times are changing so fast in technology that the information I wanted to share seemed like "old news" or irrelevant.

I have decided that I am going to release this edition anyways. Analog or digital, I think there is still much to gain from this book. There are dozens of bands and sound engineers using simple basic audio gear so I think this information could be very helpful to those interested. Digital users as well.

The principles in this book remain the same for analog or digital even though the way we do certain things have changed. Keep an open mind and you will soon see what I mean. To this day, I still make mistakes and learn something new at every show I mix. Sound is a lifelong journey. Embrace it!

Thank You for your interest!

CONTENTS

INTRODUCTION		4
Chapter 1	WHEN TIME IS UP – 3 BEST STEPS	9
Chapter 2	EXAMINATION & VERIFICATION OF ALL THE PARTS PLUS POLARITY	21
Chapter 3	POLARITY	24
Chapter 4	SETTING UP THE CROSSOVER	36
Chapter 5	GAIN STRUCTURE	44
Chapter 6	BALANCING & TUNING THE SYSTEM RESPONSE	53
Chapter 7	TUNING SYSTEM USING THE AUTO EQ WIZARD	68
Chapter 8	DUAL FFT	89
Chapter 9	FINAL THOUGHTS	118
Glossary		120

INTRODUCTION

Welcome Sound Gurus! Hopefully soon right? I have an incredible amount of information to share with you on **HOW TO MAKE YOUR BAR PA KICK BUTT!**

I sat down one cold winter with lots of coffee and started typing like a mad man on my experiences as a front of house engineer. I basically went from not knowing a single thing about sound system PA's to mixing many of the "Big 80's – 90's bands from the past in about 7 years.

In the sound industry mixing for only 7 years is considered "green". It takes years and years if not a lifetime to master this stuff. There are several factors that helped my journey immensely:

1. The fact that I am a musician (guitar player) that has been playing for close to 30 years now. I have developed an ear for good guitar tone and also for what other instruments should sound like. This helped big time! I also knew when a PA sounded less than optimal & frankly, many of them did.

2. The extreme **passion** I developed for sound once I purchased my first PA and started diving in. Man, did I discover some things other local sound guys were doing wrong. Don't go down that path!

3. By NOT trying to re-invent the wheel and I found myself a mentor to coach and teach me what I needed to do.

Here is a quick background on myself: I have a science background. I went to college and received my bachelor's degree in Human Biology. I then furthered my career by receiving my doctorate in Chiropractic. I practiced Chiropractic for about 13 years but music just kept pulling me back in that direction. Here I am running my own practice, playing guitar in a band and then learning how to run a PA… life got crazy! My passion for "Live Sound" grew and the next thing I know I am being asked to mix for bands I watched on MTV growing up. Something had to give... my passion for live sound became stronger than Chiropractic. So I stepped away from the profession.

My friends call me "Dr. J". It isn't something I came up with on my own but my friends call me that all the time so I accepted it. I bring this up because of my training in the science and the health field. I guess you can say I like to "dig in" and understand the WHY of everything.

For example, right after I purchased my first rendition of a PA, I went and watched a popular band perform an outdoor show and I positioned myself behind the FOH guy watching his every move to try and pick up that "Gem" that was going to give me that edge.

What I discovered was a sound guy who had his head buried in the graphic eq. He never paid attention to the band, never boosted any guitar solos and the sound got worse over the duration of the show. I was perplexed to say the least... how disappointing to see what I knew was a great band performing, knew the PA system was of great quality (EAW) and it still sounded like crap.

I thought to myself, "What is this graphic EQ's role in deserving so much attention that the sound guy NEVER looked up for the entire show"? I was fed up! I have seen this repeatedly over the years and it was the same thing over and over again. There is way too much focus on the graphic EQ... So I used my desire to know and "dug in" to this device.

It didn't take me long to discover that the graphic EQ was simply the wrong tool for the job. A 31-band EQ on the MAIN OUT ruling over all the channels on the mixer? Everyone is taking a beating because of this thing. A cut on the graphic makes **ALL** channels pay for it. That doesn't seem fair. I knew there had to be a better way.

As fate would have it, my analog crossover took a dump so I had to buy a new one. My brother-in-law had his own PA and told me I needed to get a "Driverack" or some sort of speaker management device because it had a crossover in it and a lot of tools to help set up the PA. I was excited because it looked really high tech. He also told me it had a graphic EQ built into it and I could get rid of the cheap 11-band eq I had.

I was glad to get rid of that extra piece of gear in the chain but still wasn't thrilled about the graphic eq in the driverack…. "There is NO WAY that great sound comes from mastering this thing!" That was my thinking. So funny looking back on this now.

With the complexity of the Driverack, I joined the Driverack Forum. This is where users can get together and chat and help one another. Some people were helpful and others made you feel like you should just hang it up.

The moderator of the forum started personally helping me. I don't know why but I think he sensed I was interested and truly wanted to learn. He said to me one day, *"I will teach you everything about the Driverack and setting up your Pa and even how to mix but you have to forget everything you **think** you know about sound"*.

I was like… SURE! Sounds great to me! At this point, I really didn't know anything about sound. I was just getting started. I didn't have any bad habits or anything like that to break unlike some sound guys.

What I learned about the purpose of the Driverack was life changing for me. The "Tool Set", the Parametric Eq's and WHY they are the EQ of choice…. NOT the graphic EQ. How to setup the Crossover; ALIGN the speakers so they fire as a team. What a Frequency Response looks like & "Tuning" the system and on and on…

Almost overnight my PA was sounding better than I have ever heard it and better than most bar PA's in my area.

When I mixed, I was able to be creative rather than being on the diagnostic side of the mixer. Remember the guy at the outdoor show with his head buried in the graphic eq? He was on the **diagnostic** side of the mixer… he never mixed. He never had the chance to be creative. He was troubleshooting. Troubleshooting the whole dang show. If this sounds familiar keep reading. Don't let that happen to you anymore.

I want to show you how to NOT be on the DIAGNOSTIC side of the mixer and start enjoying the shows you mix.

This isn't going to be easy. It is going to require **dedication, listening, researching, experimenting…** possibly breaking old habits & taking action. I can help you get up and running faster than ever if you follow my advice. Learning from those who have done it or ARE doing it is a great way to avoid costly mistakes by trial & error.

Are you ready to **MAKE YOUR BAR PA KICK BUTT?**

DISCLAIMER:

As the title states, **"How to Make Your Bar PA Kick Butt"** is just that. This book is NOT about "Mixing" necessarily although I will probably talk about it some throughout this book or at some point in a future book. It is about how you get your PA at the best possible **STARTING** position so that you **CAN** mix. Huge difference! The mixer is where you get to be creative, rather than using as a tool to diagnose and fix speaker issues.

This information is **NOT** for the guys doing arena shows where there are "Line Array" PA systems, side hangs flying, delay towers, multiple sound techs, hundreds of thousands of dollars worth of high tech gear and all the toys & gadgets you see from the big boys. They already know what I am about to tell you. They are already using gear where the engineers worked out all of the system issues. That is why good speakers are so expensive. A lot of technology went into them. So this book isn't for them, but for you, the new sound enthusiast who wants to make his or her PA sound the best it can without spending thousands of dollars.

My goal here is simple: **TO HELP YOU SOLVE PROBLEMS** with your basic PA that you may encounter at every gig, but aren't quite sure how to do it.

However, in order to get there you have to start somewhere like the beginning. We simply have to do our best with the gear you have. That is what we are going to do here and the variables are never ending it seems.

Just keep in mind… EVERYTHING DEPENDS when it comes to live sound & you **WILL** get better at this.

I would like to pass this information onto you, the **LOCAL** guy or girl who has a basic PA system on some stands…. Maybe a sub or two on up to a medium size system of a few boxes per side and SOME subs that can deliver quality sound to a nice crowd or maybe a few hundred people.

The scope of sound is huge and instead of writing a book that takes you through the history, endless terminology & advanced topics in a particular order, I would like to present the concepts to you in a way that are simple & applicable. With that said, we are going to jump into this with the first topic of discussion on **"WHEN TIME IS UP – 3 BEST STEPS"**.

Thank you for your interest in **Making Your Bar PA Kick Butt!** Dig in!

-Dr. J

WHEN TIME IS UP – 3 BEST STEPS

"Cut to the Chase - it's Showtime"

Let me cut to the chase here…. I try to frequent shows when I am not mixing and I am often asked, "Hey Jerry (my name is Jerry & my friends call me Dr. J), what do you think? Does everything sound okay? Or what would you do next?" I often cringe at this question and many times it is simply a matter of me saying, *"We are too late in the game to address these issues now"*. This is not what they were hoping to hear but all too often, it is the truth.

Unless you own a newer, more expensive PA system, the chances are high that your knowledge of knowing how to set your system up becomes crucial. In other words… if the system is an older non-powered (passive) system, you absolutely need to know how to set the system up. If the system you are working with is powered (active), chances are the manufacturer has figured out many of the system issues that come with speakers since the crossover network and power supply are contained within the box. I hope that makes better sense.

There are many factors to consider:

1. Proper Speakers & Sound Pressure level required for show
2. Proper power available for speakers
3. Verification that ALL components are working correctly
4. Polarity of all components is correct
5. Crossover setup properly
6. Gain Structure
7. System Frequency Response balanced & set correctly & ALL speakers Time Aligned. Tuning the PA.
8. Speakers aimed and placed correctly
9. Boundary & Room mode compensation & ringing the system out. Literally dozens of factors at play here.

So you can see that without me or you knowing for certain that these requirements are met, I have no choice but to say, "It is too late in the game at this point". Don't get me wrong though.... If it sounds good, it is good. If they want some tweaks, I may make some suggestions.

Like so many other things in life, preparation is the key!

Many of you probably don't feel like reading a bunch of technical stuff at the moment. So I want to give you something to try immediately at your next gig. Keep in mind however, after the gig, we will need to take a step back and start at the beginning. There is a lot to do.

What I would like to do here is list a few things that I think are critical to getting you started in the right direction. This is in NO way an exhaustive list. In fact, IF there are issues going on with your sound system, you could experience NO change or possibly make the sound worse. This will all depend on if you try this during a live gig & if you set up your system correctly to start with.

Over 95% of my prep work has been done in my garage and backyard **BEFORE** going to the gig. This pre-gig preparation will hopefully take care of the system's issues. That way when you get into a room you will only have to deal with the room issues (what the room did to your system) and not the System + the Room issues. Otherwise, what caused what?

There are an incredible amount of factors to consider (a VAST checklist) and we cannot cover them all here. This guide is for those that are **out of time** and have to make the show happen. It is a terrible way to approach a show that can make or break your performance & credibility BUT if you are out of time and have to make the system work, this is what I would do:

My #1 FIRST move is to LISTEN to the system with music. Hear me out on this.... Take an album that has been recorded well and play it thru the PA with NO Channel EQ "On" (if you use channels to play music) and NO MAIN OUT EQ "On". If you have a speaker management device like a "driverack" and you have eq-ing going on inside that device you may have to turn off those eq's as well **ONLY** if they are **NOT** official factory tunings.

If they are tunings that you came up on your own, then simply toggle them "on & off" as you go through this procedure and if they sound better "on", you may keep them. If they don't, we need to change them. The point here is, **music should sound good through the system with no board eq on**. Often this means the speaker management device has to help get the sound system to a place where the board is neutral yet sounds good with no eq.

Ok, turn off the channel strip EQ and the Main out EQ, possibly driverack EQ and listen. The idea here is that the studio recording, producing & mastering engineers have spent an incredible amount of time at capturing & delivering a product that transfers to a variety of listening devices. In other words, their intentions are for you to NOT have to EQ the track when you get it. Not every listening device has an Eq on it. It should already be in the ballpark. Make sense?

With that said – music playing through your PA should sound decent with NO mixing board EQ on it. **Think about it.** Now, if music sounds thin and harsh through your system, you probably either have too much high end in the system or perhaps no mids or subs. If it sounds muddy and unintelligible, you have too much low end or low mid frequencies going on. We usually don't say "Wow, whoever recorded that song didn't put any highs in it" because we know the system could be at fault rather than the recording. **For this exercise, assume the recording is fine and your system has some spectral balance issues.**

There are several ways to balance this all out but here is where things can get very tricky. I won't go into crossover settings in this quick guide BUT it does play a very important part of good sound. Seamless sound. Hopefully you have your crossover set up correctly. I cover crossovers more thoroughly in the section called… "**SETTING UP THE CROSSOVER**". I go into more detail there.

Ok let's continue….. The System Processor or "Driverack" or DSP as they call it is for dealing with the Speaker's DIRECT field flaws that EVERY speaker has. The DIRECT field is the sound you hear coming out of the speaker BEFORE it hits any reflections. It directly hits your ears first.

Parametric Eq's are used to correct this direct field because they can be placed exactly where they need to be. Output (post) crossover parametrics are used for correcting direct field flaws. A lot of DSP's today have INPUT parametrics as well. So the output parametrics take care of the direct field flaws inherent in every speaker. Also called "tunings". Once these are in place, rarely require adjustments. Especially, if they are factory settings. If they are factory settings, leave them alone!

Next, the input parametrics are just like an outboard parametric in between the console and crossover. In many cases, a graphic eq is placed here. So the input parametric or graphic eq can be designated as "Venue EQ". The main out eq on the newer digital boards falls into this category.

If you do not have one of these "driverack" type devices then I hope you at least have a graphic EQ on the main out bus or on the board itself. If you don't, this will get really difficult to do.

Again, the MAIN EQ on the console or outboard (graphic or parametric) is generally used to set the **tonal** balance of the system and handle Room Modes at different venues. You will want to use this EQ to set the system balance **BEFORE** you start using the Channel Strip EQ. **VERY IMPORTANT!** Equalization does have an order to it.

You could just use your ears and balance out the PA using the Main EQ until it sounds good to you BUT that may be a bit subjective so let me encourage you to use a decent set of headphones to help you with this. The headphones become a guide or a **reference** in the same way test signals (pink noise) become a guide or reference.

Here is how you do it: Take a set of headphones that have a known flat or smooth response. I use Ultrasone HFI 680's but you could use cheaper ones and still achieve great results. I wouldn't use headphones that have a ramped up low end like Dr. Dre's Beats headphones. Go with smooth and flat response headphones if possible. Listen to music thru them on your CD player or computer. Try to use an actual CD track instead of an MP3 if possible.

Have the music going thru the PA as well. Hook up the headphones in a way that you are hearing the track with NO EQ on it either by plugging straight into the headphone jack on the CD player or the headphone jack on the Mixer. If you use the headphone jack on the mixer, make sure the changes on the Main EQ do NOT affect your headphone mix. You don't want that. That will defeat the purpose of this exercise. You are using a known track with headphones as your guide to set your PA system response.

Set the volume of the PA and the headphones so they are equal in volume. Listen with headphones ON your head (exactly the way the mastering engineer planned with NO EQ) and then take the headphones OFF and listen to the PA. Do they sound the same or are they different?

Go back and forth with headphones ON and then headphones OFF while adjusting the MAIN OUT system EQ until the PA system sounds similar to the recorded track you hear when the headphones are on.

This is also where you can apply EQ to the DSP device or driverack instead of the main out eq on your board if you have one. Again, use factory tunings first if you can. These are always located on the outputs of the DSP. If no factory tunings exist, you could use these available parametrics instead of the main out eq on the board or outboard eq. By doing so, you free up the outboard or board eq to handle other things like venue / room mode issues.

So the Main Out EQ changes the tonality of the PA but should NOT change the tonality of the mix in your headphones. If it does, try hitting the **PFL** or **SOLO** on the **channel** with music playing so you can hear the music pre-fader & before any eq is applied. Then select **MAIN EQ** to make tonal changes. It shouldn't affect your headphone mix but **should** change the tone of the speakers. Make the speakers sound like the headphones. Again… music channel is in PFL or SOLO mode so you can hear it un-eq'd, Main eq is selected to make adjustments or just reach over and adjust your outboard main eq.

A decent set of headphones that have their own frequency response used with a recording that has a great transferable response can be used as a guide to help you set your PA response. It takes practice, but you can do it. It is how a lot of engineers do it on festival systems. Assuming of course every other detail of the system is setup correctly by the sound company. It is easier for them to do this rather than play pink noise. Engineers can be setting the tonal balance of the system while recorded music is playing and the people don't even know they are doing it. Pink noise will make enemies real fast... Lol

One last tip on this: you may have to shut the subs off at first while doing this with headphones. Depending on the type of live music being played usually depends on the amount of low end buildup required... i.e. Heavy Metal band or a Rap concert. Start off with the subs off so you can hear the Top PA speakers better and once you get a close match – add the subs back into the mix. Take note of low end room modes where the room wants to hang onto certain notes longer than others... tame them down with EQ accordingly.

Keep in mind at Live Shows we may want to exaggerate the low end a bit more than the recorded track but we also need to make sure the room modes in the low end don't hang on so long that it MASKS over the top of the other notes being played in the low end.

Be careful to not boost the high end too much. On recorded music we usually hear a crisp high end. It will be hard to match that up & even if you are able to, you may not want to do this in a poor acoustic environment (concrete club with a metal ceiling). This can lead to a lot of high end issues. Don't over boost this area. Also, don't be afraid to CUT this area as well.

If you are successful at this very big FIRST step – you can then begin to use the channel strip EQ to EQ your instruments.

#2 – Where are the Speakers pointed?

Way too many times I see speakers pointing in weird directions. My advice is to keep the speakers OFF of the walls and onto the people. Simple as that! You will never be able to please every seat in the house because some seats aren't even close to where the speakers are pointed. So point them in a manner where it covers the most people who are *interested* in your live event but yet keeps the sound off the walls. Done! Some speakers allow you to TILT them downward a bit too. Even better.

#3 – Turn it Up! Ringing out the System

This is where you are going to bring the system up to show volume. If you achieve show volume with NO feedback… Excellent! Have a great show!

If you start getting into feedback you will have some more EQ work to do.

Knowing **WHAT EQ** to use is where most sound guys get it wrong. I will present some options here.

1. We generally have the system processor EQ (DSP) – this is for DIRECT field correction of the speaker & box combination. This is where the word "Tunings" comes from. They live in this device and once applied, theoretically never get changed. Parametric Eq's Rule!
2. Main OUT EQ – this can be on the mixer itself or externally connected after the main out on the mixer before it goes to the crossover. This EQ is for setting the Tonal Balance of the system in different venues. Mostly boundary and room mode corrections. All too often the EQ here is abused. Historically, this eq has been used for feedback control. We need to change this mentality to last resort only.
3. Channel strip EQ – used to correct the source / mic combination.
4. Auxiliary EQ – Eq's on the Auxes to control the monitors.

Depending on what EQ's you have available will depend on how we approach this. If you have all 4 Eq's available, we should be able to do a great job here. If you do not have all of these available, we will have to do our best. It is all we can do right?

After system is adjusted for the venue using headphones and music, move onto getting the monitors loud & clear. Be mindful of your microphones. Are they Cardioid (sm58) or super/hyper Cardioid (beta58)? If you don't know you need to look this up in the microphone operation manual.

The type of microphone dictates the monitor position. Straight on or monitors positioned at the 10 & 2 o'clock position? It does make a difference because of the pickup pattern of the mic and the rejection zones.

Bring the volume of the monitor up. Do you have a way to stand in front of that monitor while you adjust it? I use the iPad for this using the app that controls my mixer. I have the monitor eq in front of me and as I push the volume I make adjustments. Some of the adjustments are in the 200-265Hz range, maybe even lower and that takes care of the low-mid "woof" stuff when you say the word "TWO". One & Two are good words to use for low-mid frequency adjustments.

At first, my goal is to make the vocal sound good through the monitor. At this stage I am tonally shaping it to make the vocal sound correct. Eventually as the volume is increased, (no singer wants more monitor do they?), I will get into feedback frequencies. I then make cuts there. The goal is to get maximum gain before feedback YET maintain clarity. So the whole process is started by tonally adjusting the EQ to make your voice sound good and clear. As you push, you get into feedback. Adjust those frequencies while trying to maintain a clear vocal. If you cut one area hard you may have to boost the neighbor area (below or above) to regain clarity back. It is a trade off of going back and forth until you have the monitor set in a such a way that it is loud, clear & stable. This takes practice.

A good iPad or iPhone RTA app is great for helping you find those feedback frequencies. Use a **medium** to **slow** mode so it will hang onto the trouble frequency long enough for you to identify it. Don't worry about the old guys who say, I do it by ear and I don't need any of that. Well, there isn't a better tool out there for you than to start using an RTA app. Letting the eyes and ears assist each other is a great way to learn. Putting a hundred people through constant feedback while you hone your skills &

track it down by ear is not a good way to make friends. All they want is for it to stop. Time is critical here.

Now we move onto the Main system.

Bring the system up to show level with the vocal mics un-muted. Hopefully you will achieve a great sound level without feedback.

If you hear feedback and aren't sure if it is the monitors or the mains, pull the fader down on the singer's channel so the monitor level is still up. If feedback goes away then it could be the Mains causing it. Try the reverse…. Leave channel fader alone and reduce monitor level. What was that result? Did the feedback go away or is it still present? If feedback went away, then it was probably monitors. If still present, then I would say the Mains are having an issue with the channel EQ or volume.

What does the EQ look like on the Channel strip for the vocal mic? Are there any frequencies boosted on this channel? Are they the same frequencies that are causing the feedback? If so, try reducing the areas on the channel strip EQ.

If the monitors are causing the feedback you simply have to go to the EQ on that monitor and ring it out some more.

You may have to make cuts on both EQ's (Channel Strip EQ & Monitor Aux EQ) to get the feedback to settle in order to get sufficient volume with stabilization.

Only take out enough to reduce the feedback and only repeat the process a few times…. Bring up volume until you hear feedback… identify frequency or frequencies … make the cut. There is a point of no return where you have cut so much and there is no volume left. So really after you repeat the process two or three times (for the mains & then the monitors), that is about all you are going to get. Know when to quit. You may just be in a horrible room with a metal ceiling. There isn't much you can do about it.

Let's say you ended up cutting a lot out of the channel EQ for all vocals around 1.25kHz... like some -12dB for example. It may be that you just have too much of that 1.25kHz in the system.

This is where you would go to the Main Out EQ and take **SOME** of it out there instead of all of it on the channel. You shouldn't have a channel eq with all the highs cut out. That tells me there is a greater underlying issue somewhere else that you will have to investigate.

Just be wise about where you do your Eqing. If you did a good job of setting the response using music and headphones, then this really shouldn't be an issue, BUT, the more volume or gain you put into the system, the chance of feedback goes up.

What I often do when I know I have a show with a "Mic Cupper" is get myself prepared by placing the singer's mic in the front center of the stage. I point the mic straight up towards the ceiling and un-mute the channel and increase the gain forcing it to feedback. I will make this cut on the channel eq. Hopefully, no more than 3dB to start with. I will then place my hand over the mic slowly to see if I get an additional screams or howls. I will make that cut as well (wherever it may be). After that I stop and go directly to the main EQ and mark those same areas. Not necessarily make any cuts but if I need to, I already have them marked as my last resort. Almost like a pre-emptive strike. If you have a vocal "sub-group" eq, you could go there before the main eq that way you don't make the entire band pay for an unruly mic. Remember, any cuts on the MAIN EQ will have an effect on everything. Also, keep these feedback areas in mind for monitor eq-ing as well.

In any case – you are striving for the best of everything here: Loud Clear Vocal with no feedback while making minimal cuts on the EQ. In the end, do what you have to do to stabilize the system and keep the show going.

Sound DEPENDS on so many factors. There is so much to all of this that many get overwhelmed and tune out because what they really want to do is grab that EQ and start working some kind of voodoo that makes the band's sound great. If only it worked that way. I hope this guide was helpful. Nothing can beat preparation before a show but like I said, if I was out of time or if you are out of time – this is what I would do.

CHECKLIST:

1. Use CD track along with headphones to guide you through setting up a Tonal Balance for the system. I use the "Dark Horse" album by Nickelback mostly because Robert John "Mutt Lange" produced the album... he did all the heavy hitters back in the day when albums were not compressed to death... AC/DC, Def Leppard, Bryan Adams, Shania Twain... many more... Good pure natural sound.

2. Point Speakers away from Walls and onto the people most interested. Try to go with a height where the horns are at least a foot over most people's head.... So 6 and a half to 7 feet high should be good. Up higher – even better.

3. Ring out the monitors first. That way when you bring FOH up - it may not react as much. Ring out monitors while retaining clarity.
Know the type of microphone you use. Is it a Cardioid? A Supercardioid? If you don't know, look it up. It makes a difference where you place your monitors.
If you use a standard SM58 – it is a **Cardioid** mic and requires the monitor to be placed **straight** behind the mic.
If you use a Beta 58 – you have a **Supercardioid** mic which states **very clearly** in the manual that the monitors must be positioned in the **10 & 2 o'clock position** for maximum gain before feedback. Know your microphones and monitor positions! Knowledge is key!

Finally, bring up the FOH and un-mute vocal channels including auxes (monitors) so the entire system is up and running. Place vocal mic in the center front stage and point the capsule straight up to the ceiling. Force into feedback slowly if you can. Place a cut on the frequency. Perhaps "Cup" the mic a bit to get additional frequencies to react. Make another cut. STOP. Do this on the channel strip eq. If feedback persists, make a few cuts on the main out eq to see if it helps. You should not have more than a few cuts for feedback control on the main out eq.

When you have a powerful singer, you will find that you won't make very many cuts at all. When you have a weak singer (whisper singer), it forces you to gain up their channel because they are not giving you the gain you need. **This is really where the problem lies.** They are not giving you enough gain from the source. Their own voice. So this forces you to deal with trouble frequencies.
Unfortunately, you will have to deal with weak singers your whole sound career so you may as well get good at this.

Keep in mind – if the system is now lifeless and sounds like it is under a pillow, then try to relieve these cuts a bit to bring life back into the system. You at least have the areas "marked" if you need to cut.
I don't always leave my cuts in place but they are at least "marked" if I have a weak singer or a mic cupper hitting the stage.

I hope this helps. Give it a shot. We will revisit this again. There is a lot to sound systems. We simply cannot cover all the topics here when it is near show time. Get yourself prepared in a **BIG** way for your next show.

EXAMINATION & VERIFICATION OF ALL THE PARTS PLUS POLARITY

I hope the last section was helpful to you. The more you do it the better you will get and you will also have to do those procedures from here on out for the rest of your sound career. So master it. Now that we got some of the "I want it now" stuff out of the way we need to dig into the nuts and bolts of a sound system.

Before we start setting crossovers, turning knobs, pushing faders & eq-ing like a mad man we need to examine all the parts of the PA. If we don't take care of this now we WILL have to address it later. So many beginners skip over this. This can be fatal for optimal sound. They buy a few speakers here and there – throw it all together and expect it to just work. You cannot effectively optimize and tune a system that is **unverified**… it will be a waste of time.

For me, it really messes with my thought processes big time! If something isn't working like I want it to, I want to be able to **RULE OUT** as quickly as possible the basics. "**YES** my speakers are wired polarity correct. YES my mic cables are polarity (continuity) correct and in good shape. YES my speakers are properly matched with exact ohm loads. Yes everything is hooked up correctly. This helps you to rule out many things so you can narrow your search when you are under the gun.

This tedious process is **essential** to setting up and tuning the PA. There is nothing like **KNOWING** your system forwards & backwards & knowing the parts match & are working correctly.

While you are examining all of the parts, make sure you know for certain the **OHM** rating of your speakers. If you have Powered Speakers – you may have to pull a speaker and horn out and look to see what the rating is and use a multimeter on it to confirm. It is the only way to know for sure. If it is a brand new speaker out of the box, don't void the warranty by opening it up. You will have to trust they got it right.

To verify passive speaker's OHM rating, use a multimeter and set it to the OHM (Ω) setting. Since most speakers are 8 ohms, set the meter as close to the 8 ohm range that you can. Now, place the probes on the (+) & (-) terminals. If you get close to 8 ohms or even a little less, you are fine.

If one says 3.2 & the other 7.1 ohms – then one is an 8 ohm speaker and the other one isn't. It is most likely rated as a 4 ohm speaker. Go through all your speakers and verify that they all match within a certain ohm range. Hopefully you will get similar readings.

Most of my speaker readings today read around 6.3-6.5 ohms even though they are rated as an 8 ohm speaker. This is perfectly normal to get a slightly lower reading.

If you determine that you have an 8 ohm horn driver on one side of the PA and a 4 ohm horn driver on the other side you will have balance issues. One will be louder than the other. This can be balanced using your amplifier input sensitivity if the driver is on its own power amp but really all speakers need to be paired correctly.

Nothing beats a correctly matched system and the last thing you want to do is start implementing all kinds of "workarounds". So we need to verify this.

If you have to make "due" with what you got – keep proceeding. We will have to make it work until you can get the matched set.

I have a spare horn driver here. (Scroll down to next page)

It is a Eminence PSD2002–16 Ω driver. A very common driver in the Yamaha Club Series speakers. Let's measure and see if that is correct.

We have a reading of 12.1 ohms. This is perfectly normal for the reading to **NOT** be exactly at 16 ohms. The ohm readings will probably never be over the rated value but they can be slightly lower.

The tool I used to measure this is a "True RMS" multimeter. There are two different types of meters in the picture. The "Cen-Tech" Multimeter literally cost about $5.00 at Harbor Freight. The Craftsman "True RMS" multimeter was about $60 at Sears. Very helpful tools for measuring and checking pins on mic cables…etc. Even voltages in the system.

POLARITY

Many of us have heard the word POLARITY before but haven't thought about what it really means.

Polarity? Yeah I am sure you have heard this word before but it is important that we get this right. This word gets so confused with **PHASE** that most people think they are the same. They really aren't. Polarity is either IN or it is OUT. It is ABSOLUTE. Zero degrees together or 180 degrees apart from each other. Positive polarity is the same as two race cars going around the track TOGETHER neck and neck. The Input is moving in the **same** direction as the output.

Why is this important? Because your speakers are either "IN" or they are "OUT" of polarity. A positive impulse on the positive terminal of your 18 inch sub will make the cone move OUTWARD. A positive impulse on the negative terminal of your sub will make the cone SUCK INWARD. Did you catch that – SUCK? It will literally… and the low end will seem distant if one sub is in polarity and the other is out of polarity. Any wonder why you can't hear someone's kick/bass drum but yet they have 8 – 18 inch subs? How is it possible? What about your mids? Horns?

If you have one sub "IN" and the other "OUT" they will cancel each other. There won't be complete silence but no matter how hard you push the subs you won't be able to hear the KICK drum very well. It will have a "LOST" confusing sound & lack punch. The same thing applies to mids and horns, especially for vocals. Take a mixed bunch of IN polarity speakers & mix them with OUT of polarity speakers and you have a disaster on your hands. NO amount of tweaking at the board or graphic EQ will fix this.

It seems crazy to think this but many times people buy speakers that have had previous owners and sometimes these owners re-wire them or do maintenance work or even replace drivers not giving much thought to the importance of Polarity. Some load cabinets up with a different brand of speaker. Maybe the speaker doesn't like that box. Not every brand of speaker will play nice with each other. One may have a peak at 3kHz and the other doesn't. Eqing affects both. Maybe one doesn't need eq'd. We can't leave this verification step to chance. I have even seen factory speakers wired wrong… maybe it was a Friday or something.

Ok – you get the idea now. Woofers are easy to check. All you need is a 9-volt battery. However, I would recommend purchasing a **POLARITY CHECKER**. It is a great investment and would cost around $100. You will use this device constantly. You can check continuity with MIC cables and the checker will tell you if the cord is good OR bad and what wire or pin is bad. GALAXY AUDIO has such a device and it is called a "Cricket". See Pic.

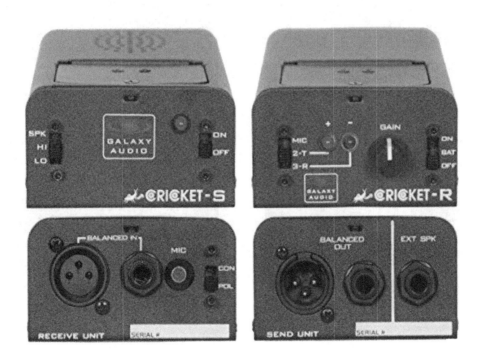

Charts for the Cricket: Makes checking Mic cables a breeze!

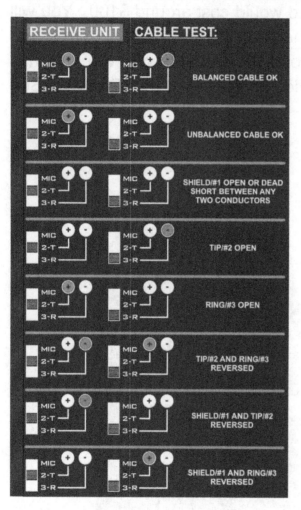
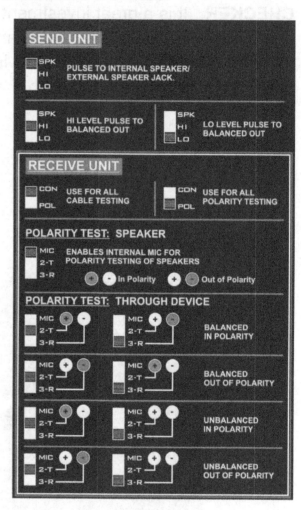

The Cricket will tell you if the polarity is IN or OUT. This will be especially helpful for testing HORN drivers. Since the diaphragm is enclosed – you can't SEE the tiny cone move IN or OUT. The cricket will tell you just by putting an impulse into the horn without ever taking the horn out of the cabinet.

It makes checking speaker boxes easy as well since you won't have to open the box up to check the speaker.

IF you just can't afford anymore gadgets right now you can use a 9 volt ONLY on WOOFER type speakers. DO NOT USE a 9 VOLT on the horn driver. It could blow it. You can check a horn driver with a 1.5 Volt double AA battery just to see if it works. It will be a "click click" sound. It still won't tell you if the horn is in positive polarity though.

Procedure for checking speaker POLARITY:

What type of connector do you have on your cabinets? Quarter inch jack? Neutrik Speakon connector? If needed, you could make up some connectors that will plug into the cabinets with their existing jacks BUT with the bare wires on the other end so you can get a nine volt on there. OR, you could take the speaker out of the cab and set it on a bench and check it there. A positive (+) impulse delivered to the positive (+) terminal on the woofer will make the cone move **OUTWARD.** If you flip the battery around (+) for the (-), the woofer should SUCK **INWARD.** If you make connectors up so you can just hook a nine volt up to your speaker jacks – make sure you visually see the speaker move outward when a positive impulse is applied. Try not to keep the nine volt on the woofer too long. Just an "On & Off" type of rhythm to see the reaction of the speaker. This will tell you if the manufacturer wired the speaker correctly (most likely) and you no longer have to question the polarity of the woofer unless you swap one out. Move onto the next connection…i.e. the ¼" jack or Speakon connector.

The first PA system I purchased was some used stuff I bought off of several different people. More or less a "hodge podge" of speakers and amplifiers. I tried mixing on this system and I knew something was wrong but just didn't know how to fix any of it. I tried everything I could think of at the mixing board but nothing helped. The embarrassment of trying to diagnose this stuff during a gig made me want to sell all of it.

I am glad I decided to dig in and figure out what was going on. Turns out I had a few speakers out of polarity with each other and one monitor out of three - **OUT** of polarity. As you can imagine, I was having cancellations going on all over including up on the stage. **Examine and Verify** everything **BEFORE** your next show. **I cannot stress this enough.** If something in your system changes like a new piece of gear, you replace a speaker, bought a batch of used cables off of a friend…. It doesn't matter what it is… you examine it and verify that it is operating the way it is supposed to. **ALWAYS!**

The next gig out with my system, things seemed much better. I had better clarity and focus of the system. Amazing! Just from getting the Polarity correct.

With the polarity checker I was able to check every single cord in my arsenal, every single speaker, every single horn in an hour or two. I even found a couple of MIC (XLR type) cables with inverted Pin 2 for Pin 3. You wouldn't think a simple swap of pin 2 & 3 could cause so many problems BUT IT DOES! After all -- microphone cords are what we use to hook our mixer up to the crossover and onto the power-amps. They have to be checked. By going thru this tedious process (some say anal – who cares what they think) you really learn your system inside and out. You gain confidence in it and when a problem arises as you know it will – you will have a pretty good idea where to start. Number one item ruled out = POLARITY. Number two ruled out = mic cables…etc..

I am thankful I had problems with my system in the beginning otherwise I wouldn't have learned what I have. I now know my system forwards and backwards. This comes in real handy if an issue comes up. You can automatically rule out things in a hurry. Polarity is NOT going to be one of those issues or a mis-wired cable.

This next chart is today's standard for wiring XLR cables. Memorize this…

> The Standard Wiring Configuration
>
> for XLR type cables (Mic Cables):
>
> Pin 1 is the SHIELD (ground)
>
> Pin 2 is the HOT
>
> Pin 3 is the COLD

Check all of your owner's manuals to make sure your mixer, crossover & power-amps demonstrate this wiring configuration (they should) & then physically check them with a cord checker / multimeter or "Cricket" type of instrument -- If you can.

Start at your speakers and check them. Make corrections if you have to. From there – check the speaker **jack & cables** (we are moving backwards to the mixer).

A couple of words on speaker cables:

They can be visually examined or checked with a "Multi-Meter". Just make sure the tip is going to the tip (the positive) & the ground is going to the ground (the negative). This would be for the quarter inch type jack connections. If you have SPEAKON connectors make sure (+1) is going to (+1) on the other end and (-1) is going to the (-1) on the other end.

For Bi-amped tops where you have horns on their own amp and mids on their own amp – it doesn't matter if you choose the "2" for horns or the "2" for mids – just keep them separate and straight. For bi-amped tops I use Neutrik Speakon connectors called an NL-4. You can use four wires on it. So my mids are connected to (+1, -1) & the horns are on (+2, -2). All on one cable (a four wire cable). If you prefer to put horns on (+1) and mids on (+2), go for it as long as you **know how** it is wired if a problem arises. It makes diagnosing way easy. If you have power-amps that have "Banana" plug connectors on them. Same idea. Positive from speaker to positive on amp…etc. One word on Banana plugs…. They are easy to invert and therefore easy to accidentally hook up incorrectly. Pay attention and mark the plug somehow that way you know the positive is going to the positive.

Now that we are working backwards from the speakers and are at the amps – we make the switch from two wires (+ & -) to the XLR style Mic cable. These cables run from the power-amp INPUTS back to the Crossover OUTPUTS. Don't try to hand make these cables unless you are really good at soldering. XLR Style mic cables have 3 pins inside them. Pin 2 is the (+) HOT Pin. Pin 3 is the (-) COLD Pin & Pin 1 is the shield.

Buy some good quality cables for this. Most high quality cables have a life time warranty on them and IF they fail will be replaced without question. Let me back up just a second here... You can also check the polarity of your amplifier IF you have a polarity checker. If you still don't have a checker yet – we can try an alternative method. I will explain in a bit.

So now we are at the crossover. Place an impulse into the INPUTS of the Crossover and observe what the OUTPUTS of the crossover read. It should be fine but you never know. These are just good exercises to do to gain knowledge and experience of all your gear. Check all inputs (L&R) and outputs (L&R), High, Mid & Low.

From the crossover we have to go from the INPUTS back to the Mixer OUTPUTS. This requires at least ONE XLR type cable (MONO) or two if running out of both Left and right outputs. Again – these cables need to be good quality. All good? We are finally at the mixer so now what? Well, if you use a snake we aren't quite at the mixer yet! YES – the snake has to be checked out. Every single jack / channel plus the AUX sends. You are probably saying, "Man, this is crazy!" Yep – but you gotta check it. The signal can get inverted and reverted (is that a word?) and inverted again by the time it gets from the mixer to the speakers to your ears! Don't leave anything to chance. You want to be the best sound guru in your town right? Let's keep going…

Ok – the snake checks out and we are finally at the mixer.

I would then run your polarity checker into mixer INPUT channel one. Get a signal (PFL) and set your trim or gain. Send the impulse thru the board, raise the channel fader where you have it routed to so you can hear the Cricket "Chirping" in the PA – yep that is why they call it the cricket. Now take the receiving end of the checker (there are two devices – the SEND unit and the RECEIVE unit) and walk up to the speakers and read what the polarity checker says. On the receiving end of the checker is an external mic that measures the sound impulses and determines if the speaker is in fact moving OUT indicating a positive polarity. Positive Polarity will give you a green light. An inverted signal will give you a RED light. You don't want any RED lights for now.

The cricket is great for testing horns and mid woofers. Testing subs by themselves (individually) is easy too but can be tricky while testing all the way thru the PA. There is a possibility of false readings. These are most likely due to operator error. I usually have to turn the mid and high amps down or off so the device can really read what the subs are doing.

If there is a strong impulse coming out of the subs because you have them turned way up - it may give you a conflicting reading where you see a GREEN light (polarity in) and then a sudden RED light (polarity out). This is due to mic overload. Simply turn the channel down or amp down to accommodate the checker and it will give you a decent reading. On the front of the receive checker is a GAIN knob. You can try to reduce that level as well. Basically, low end information can easily overwhelm the mic on the checker. Try using the right amount of level to get an accurate reading.

Now you need to do this thru EVERY channel on your mixer. Go ahead and do it at least one time so you know beyond a shadow of a doubt that your current setup Polarity is correct. We can't leave it up to the factory workers or robots to get this right for us or anybody else for that matter.

Now if you have ACTIVE or POWERED speakers you can still do this – you just have less cabling to deal with. We are almost done but guess what?

The Galaxy Cricket can also check polarity on Microphones! There is a procedure for that too and we HAVE to check them. Yes, microphones get wired backwards too! It would be a complete waste of time to go thru all of this and have 99% of the system in perfect polarity and still can't figure out why the kick drum or vocals can't be heard or sound weird. A reversed polarity microphone will screw it up.

Think about it: When the drummer stomps his kick drum – a positive impulse must travel from the kick drum mic thru all the cabling over to the snake out to the board and into the kick channel thru the mixer and out of the mixer back thru the snake into the crossover back out of the crossover into the power-amps and from the power-amps to the speakers and from the speakers delivering a positive forward cone motion of acoustic sound out to your ears! Multiply this by every channel on your mixer. Guitar, vocals, bass, keyboards…etc.

Now, I said earlier that I would tell you how to check all of this without a polarity checker. It is much more difficult though. Have your drummer stomp on his kick and go up to the sub and see if you can visualize with a flashlight the speaker moving forward or outward. For a vocal mic – just cup it and slightly (carefully) pop the top (poof) to see if you can see the speakers moving outward. Horns just can't be checked without a polarity checker. A good polarity checker is the way to go. It will be the best hundred bucks you spend next to a measurement mic. Guaranteed!

I hope you are beginning to visualize your sound systems signal path and why it is important to have all the components working properly. Now that everything has been examined & corrections applied (if needed) we can have the confidence that the system has been **Examined & Verified.**

So at your next gig if there is an issue, you can be rest assured that it is not a polarity issue. You can then move onto the next set of possibilities.

Final thoughts on Polarity….. There are times where we want to purposely INVERT the Polarity on a channel. For example using a mic on the bottom of a snare drum or any drum you mic from the bottom. Just make sure you know where your system is at "in polarity" to start with and if you need to invert a channel - go for it.

If you question yourself on this just experiment and use your ears… "Hey drummer – will you please give me your snare drum"? Drummer hits snare in a consistent fashion and you listen for a bit then FLIP polarity and listen some more. Did the snare drum get quieter or louder? Fuller sounding? Go with what sounds best to your ears. I would go with the fuller sounding tone. Try other channels like the kick drum and other instruments… even vocals. See if you can hear the difference and what sounds better to you. I don't have to flip polarity much except the typical snare bottom.

Always let your ears be the final judge on this.

In time, we can discuss more uses for purposely inverting polarity. Using subs clustered together in the center of the stage is a big one. Dealing with the large bass buildups that occur can be bothersome to the musicians on stage. The solution? Cardioid subs. One solution may be to "Invert" the polarity of a center sub along with some electronic delay. This may clear the stage of unwanted heavy bass… but let's talk about that more later. You will find the more you learn, the more you don't know about sound.

For now, let's build solid foundations for good sound. Correcting Polarity issues can have a dramatic effect on your system and instantly improve the sound and performance. Don't be that guy on the diagnostic side of the mixer… you cannot MIX your way out of a poorly set up sound system that hasn't been examined and verified. What we are eliminating here is "System" issues that the mixer cannot fix.

Sorry to put you through this exhaustive, possibly boring, compulsive procedure but if you got through it and everything is great now… Congratulations! You no longer have to wonder what is wrong with your system from this technical point of view EVER again UNLESS

you add something new or fail to check something you re-wire or you replaced something. But we have so much more to talk about...

Final thoughts and tips:

In the picture below I discovered that the Polarity Checker "Cricket" did not have any Speakon Neutrik connectors on the unit. To get around this limitation - I made up some various cables that would allow me to connect to the Cricket and also to my speakers. In the picture, I made many cables to help with different scenarios. Check it out…

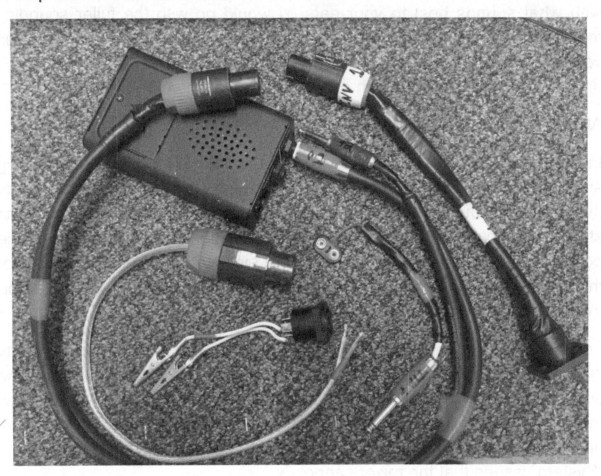

The far right cable is a Speakon NL4 connector on both ends – male and female EXCEPT it is purposely wired INVERTED. For example, if you find you have a speaker that is inverted and you place this in the path – it would flip the polarity back to its correct position. The Cricket could confirm this application. This cable could also be used on the fly if necessary BUT what we are trying to do here is prevent that from happening by checking everything ahead of the gig.

The cable attached to the Cricket here in the pic, is a MUST have. Basically, since the Cricket has ¼" connectors on it (wish it had speakon) we have to make up a cable to go from a ¼" jack to Speakon so we can get it into a speaker cabinet. For the older speakers – a standard 1/4" to 1/4" speaker cable will work great. I used an NL4 so I could wire up a (+1, -1) & a (+2, -2). This is the most important cable I have and will be yours too. Wire it up correctly through visual inspection. It can check (+1, -1) type of cabs and also a (+2, -2) type of cab. For example, a bi-amped top where the horn is wired to the (+2, -2).

The remaining cables allow me to check a (+1, -1) using an NL8 (8 wires) connector. One other cable allows you to check polarity of speakers using a 9 volt battery. I don't actually fasten the battery onto the connector since I want to do a quick ON/OFF rhythm to watch the speaker movement.

Finally, the last connector in the pic has some clips on it to check a speaker sitting out on the bench. It is used in combination with the MUST HAVE cable.

You will quickly use your imagination to design and make up whatever cable you need to get the job done. Buying specialized cables can be very expensive.

That's it! I hope this chapter serves you well. Once everything has been examined, verified & corrections applied if needed, we get to set the Gain Structure & TUNE the system!

www.galaxyaudio.com

SETTING UP THE CROSSOVER

I remember it like it was yesterday…. all the fooling with the graphic Eq & the crossover. If something wasn't sounding better using the graphic Eq then all of the sudden, further tweaking had to be done at the crossover. I saw this so many times. When I see this today, I just cringe. Especially during a show.

With the current boom in powered speaker sales, a lot of the system setup procedures have been taken care of by the manufacturer. Plug & Play more or less. So for some of you, setting up the crossover won't be of much interest if it is already done. For those of you who do use a crossover, we have a lot to talk about.

Some of you may just want to learn more about them anyways so you too can help others set up their systems to reach a higher potential.

Let's begin:

A **Crossover** or **Spectral Divider** is an electronic device that divides the audio signal into different frequency bands. These frequency bands go to different speakers such as a horn, mid woofer or a subwoofer.

The crossover makes sure the lows get the lows, the mids get the mids and the highs get the highs. The end result is to make these separate speakers act as if it was one giant speaker & to join the waveforms back together in the acoustical world in a seamless fashion… at least we hope so.

With a High / Mid powered speaker, this is already done by the manufacturer BUT we may still need to combine that High / Mid with a subwoofer. So an onboard or external crossover will be needed. Some of the more expensive powered speaker companies have switches on the boxes to act as a High Pass or Low Pass filter for speakers, effectively acting as a crossover between band passes. Take a look to see if your speakers have this function.

Most Bar PA's out there are 2-way systems. This means they have some Full Range Tops and some Subs. The Top boxes have crossovers built into them joining the horn to the mid woofer. You may have speakers that are 3-way or even a 4-way system.

Choosing the proper external (outboard) crossover is important. They make them in all shapes and sizes… a 2-way stereo, 3-way mono, 4-way mono… 2 IN 4 OUT, 2 IN 6 OUT…etc…

Make sure when purchasing a crossover that you get the right one for your needs. I would suggest the crossover device that your speaker manufacturer recommends but that may not be feasible financially. In that case, a more affordable speaker management device like the DBX Driverack PA, DRPA2, DR360, Peavey VSX26, or even a Behringer system processor can do wonders. Not just because it is a crossover but because of the important "Tool Set" available.

They act as a Crossover, they have "Time Alignment" available that allow you to time align your tops to your subs, Parametric EQ's so you can TUNE the speakers, Graphic Eq's & Limiters & a few more tools to really dial in your PA system. The Driverack has many tunings available for speakers. You could check to see if your speakers have a factory setting in the presets of this device and you may just be off to a great start with minimal effort.

There is literally anywhere from 3-6 pieces (maybe more) of rack gear in one single space Driverack. This reduces the amount of gear in the signal chain which can make setting up the Gain Structure much easier.

This type of management device has even spread over to amplifier makers where they started incorporating the tool set into the amps. Therefore, reducing more gear in the signal chain.

It is a great time period for Live Sound because the technology is getting crazy! I will try to remain focused here and make sure I keep it simple for the most basic sound enthusiast starting out. Even if that means you are using a Graphic EQ as a crossover. I would recommend against that but if that is all you have & the show depended on it, I guess I would use the graphic too.

I am going to list some basic settings for a crossover as a good starting point BUT if the manufacturer for your speakers say the crossover point for the horn to the mid woofer is 1.8 kHz, then I would do what they say.

I have tried over the years to come up with my own crossover points for my system and I have not been able to do a better job than what the manufacturer came up with. I just don't think it can be done. They have the experts on staff, the facility, the equipment, the room (anechoic chamber)…etc… so follow their recommendations. You don't want to void the warranty if you are not following their recommendations.

If you have a system that is fairly old and you cannot find any crossover data for them or system tunings, then we will have to do our best. I have listed some basic settings.

This can vary a bit depending on the type of speakers you have and what they are capable of doing. I will try to address the variances.

BASIC CROSSOVER SETTINGS

Subs: 40Hz BW 18 – 100Hz LR24
Mids: 100Hz LR24 – 2K LR24
Highs: 2K – Out

These crossover settings are NOT absolute but are safe. I used these exact same settings for quite a while with great results until I had the tools to really optimize my system utilizing a Dual Channel FFT program. I will explain what that is in another chapter.

On the low side, the 40Hz could be raised to 45Hz or even 50 Hz. Especially if outdoors. **Keep this idea in mind:** When the sound system is indoors, you have walls to trap the sound which makes your system sound louder than it really is. The low frequencies buildup and you can take advantage of that to a certain degree, minus room modes of course.

Outdoors it is the opposite. You tend to push the system harder because there are no walls to contain the sound. It may be harder to hear the low end the way you are used to hearing it inside. This is where raising the high pass on your subs can save your subs from failure due to over excursion & heat. It won't let you hear the low end better but it will help to protect your speakers better. True 50Hz is still low.

If your subs have 3 inch voice coils, this would be a really good idea. It wouldn't take much to blow a set of 3 inch voice coil subwoofers. A name brand of 4 inch voice coil subwoofers can handle much more punishment than the 3 inch. You may be wondering what is BW or LR on the crossover. BW stands for Butterworth and LR stands for Linkwitz-Riley. These are names of the type of filters that are commonly used on crossovers. Named after their inventors of course. The numbers refer to the slope. A BW 18 has a slope of 18dB/octave. The LR 24 has a slope of 24dB/octave. With these settings, you will be in good shape. If you want to learn more about BW & LR type slopes, "Google it". It is pretty boring to read but at least you will know why they came about.

Some speaker physics:

When selecting crossover points it is important to note that mid-range speakers will "beam" if the crossover is set to high. This should prevent you from setting a crossover point in that range due to beaming. **Beaming** is where the frequencies get real narrow and pierce like a laser beam. Some people describe this as an "Ice Pick" harshness on the ear drums…

I have some data here that I wrote down on a piece of paper years ago but I cannot find the source. I am pretty sure it was from a magazine or newsletter article. I wish I could remember so credit can be due to them.

The Theoretical Maximum Frequency Before Beaming (Hz) on a speaker:

5" = 3,316 (Hz)

6.5" = 2,672 (Hz)

8" = 2,105 (Hz)

10" = 1,658 (Hz)

12" = 1,335 (Hz)

15" = 1,052 (Hz)

18" = 903 (Hz)

On a recent system optimization job I was working on, I found a 15" mid woofer was crossed at 3kHz to a horn. When I asked why that crossover point was selected the guy said it was because a friend told him that it is where it should be. No other reason apparently.

Looking at the chart, it is evident that once the crossover point went above 1,052kHz or basically 1k, beaming kicked in. I then looked at the horn and the horn was recommended to be crossed at 2kHz.

Since the 15" woofer would be beaming pretty hard at 3k (like a laser beam for sure) I lowered it to 2kHz. Not as low as I would of liked but then I would risk damage to the horn if I lowered the horn too low. So a trade off had to be made. The horn is crossed where it should be and the woofer isn't being asked to reproduce anything above 2kHz.

Imagine a horn and a mid woofer delivering sound to the audience. Each component is basically "Spraying" the sound out kind of like a garden hose nozzle spraying water. Where you really want the crossover point to be is where both drivers have an equal spray pattern… make sense? With the mid woofer at 3k, the beam would be very narrow. A narrow spray. With a lower crossover point, the mid woofer's spray pattern seems to match up better with the 2 kHz on the horn at that crossover frequency.

If you have a beefy horn you can go lower and that would really help with mid woofer beaming. If you have weak horns you shouldn't go below their recommended crossover point otherwise you will damage them.

Now for the 100Hz Crossover point:

If you are going for a complete equal level between the tops and the subs then 100 Hz will be fine. To get the warmth and big low end feel that many of us enjoy from a rock show, the Subwoofer level has to be up there by 10-15dB. They call this a "Haystack" response.

When the system response is setup this way... see the chapter on **"BALANCING & TUNING THE SYSTEM RESPONSE"**, the Acoustical crossover point is not at 100Hz anymore. It is more like 150Hz because the low end buildup shifted the acoustical crossover to the right.

You have **Electrical Crossover Points** and you have **Acoustical Crossover Points.** Electrical Crossover points transfer well to the acoustical world ONLY if there is Equal Level between the two components. Once one component is louder in amplitude than the other, for example subs, the electrical crossover point may be 100Hz but in the acoustical world (the only world that matters) it may be much higher.

How I deal with this shift is purposely put a GAP in the crossover points. Tops are 100Hz and the subs are around 80Hz and this yields me an Acoustical crossover point around 116-120Hz.

In addition to setting crossover points, time alignment comes into play. You can have the right crossover points set but without proper alignment, you will not have proper phase alignment or "Summation" across the crossover points. The end result will be a "dip" in the response of the system that cannot be Eq'd and it won't sound as tight or focused in the low end.

You can do the poor man's way of aligning drivers by playing a test tone that is equal to the crossover point... for example a 100Hz test tone for subs to mids or a 2kHz test tone for mids to highs and adjust the alignment delay for that output until you hear the loudest tone because of summation.

You could also invert (purposely invert polarity) on one or the other signals and listen for the most DECREASE in volume of the tone while you adjust the delay time. It will probably be easier to hear this method better than the other but try both if you like.

If you cannot get a noticeable change in volume with either of these methods try to apply the delay to the other speaker instead. It may be that the first speaker you applied delay to was the sub and the sub didn't need the delay but the Top did.... it is hard to do this without a dual FFT analyzer.

In my experience, if you have "Front" loaded subs and Top front loaded speakers sitting directly on top of the subs, the Tops will need the delay and not much. I would be surprised if they were delayed more than 2ms. Mine are delayed 0.75ms which isn't much and I could probably leave the alignment delay off and no one would notice.

When I hang my tops up high and fly them over the subs, I have to readjust the delay to get them back in alignment.

As you can see this can get complicated in a hurry... don't sweat it... just keep trying.

When you get to the level of using a Dual FFT program like Systune or Smaart, you will be able to nail the alignment with high precision. It is expensive and requires training to read the data.

For now, the manufacturer's recommended settings are best and if there are none available due to you having some older speakers, go with the crossover settings above and use test tones to align the drivers the best you can.

You can find these test tones online pretty easily. Search for "Test Tones" & "Pink Noise".

In Conclusion: I hope this was as helpful to you as it was for me. Learn your system processor well. Sit it out on a table with the manual and start pushing buttons and turning knobs. You are not going to hurt it. I did this for about an hour or two when I first got mine and everything started to make more sense. I know it will for you too.

Enjoy! That is it on Crossovers & Time Alignment for now. We can only go so far with this here unless we get into Dual FFT, which I plan on doing in a later chapter.

You may be wondering about "Where do I set my output gains on the crossover"? For now leave everything at "0" or Unity. If you need more gain in the sub region, we should go for the Sub Amplifier. If Sub amp is wide open and there is no clipping or danger of hurting the subwoofer speakers, we can then increase the gains on the output of the crossover for the subs. Check and see.

GAIN STRUCTURE

There are two types of Gain Structure in Sound Systems: Mixer Gain Structure & System Gain structure. They both work hand in hand. Nothing beats a solid Gain Structure. Relying on Limiters to protect the system in the event of clipping can be a risk. It all depends on the limiter and how you have it set up. Some limiters just aren't fast enough and offer little protection. The best way to protect your system is to have a solid gain structure first and then limiters second.

Basically **Gain Structure** is where we calibrate the GAINS of each device in the signal chain to operate in its optimal range (best signal to noise ratio) plus set up the system where the meters on your mixing board accurately tell you what the meters on the rest of your gear are doing. So as you approach clipping on the mixer, you know you are approaching clipping on the next device in the chain and so on all the way to the power amps.

When the gain structure is set up correctly, the mixer's **Main Meters** are the only meter you need to watch since all other meters & clip indicator lights will be calibrated to this main meter. Keep in mind that just because the mixer is NOT clipping, doesn't mean the next piece of gear in the chain isn't - so let's fix that if that is an issue for you.

There is controversy over how to do this the best way so we will go over a quick "down & dirty method". It works pretty well actually.

The big differences in how people do gain structure (in my opinion) stems from their mixing habits and how they run their gains on the faders. One person may strive for a "Unity" mix approach and the other a "PFL" every single channel to read meter 0dBu or -18dBFS approach, which in turn can affect the way the gain structure is set up.

It can be confusing so hang in there. When it comes to Live Sound, you will quickly notice how it is next to impossible to get the experts to agree on many things. The truth is, it all depends and both sides may have valid reasons. One or the other may be right according to their situation. Don't get caught up in it or let it bother you. Do what is right for you. I will explain as we go through this. If you haven't set up your crossover points yet, this may be a good time to do that. **Always go with the manufacturer's settings if possible.**

So how do we do this Gain Structure thing? We will use a test signal to set this up. Our test signal for this will be Pink Noise. Pink Noise is basically continuous random noise having equal energy per octave across the audio spectrum. It has more low-frequency content than "White Noise". So, we will use Pink Noise as our KNOWN source. When we use KNOWN sources (pre-calibrated precise signals), we can often find other unknown variables. Test Tones & Pink Noise are often used in the audio field.

Use a cable that will allow you to connect from your computer, CD Player, or tone generator, etc… to your mixing board. Connect it into an input channel on your mixer. Use the necessary connectors to get the signal into your mixing board. If you do not have a pink noise audio file, you can do a Google search for it. I use the "Bink" or "Binkster" Test CD. It used to be FREE & downloadable but it may not be available anymore. The File is about 20 minutes in duration. However, if it isn't, here is a link of a downloadable pink noise file that does work.

File Location: http://www.nch.com.au/tonegen/index.html

You may also have a mixing board that has a pink noise generator built into it. That is even better.

Before we go through the procedure I am going to call the 1st. method the "Spec Sheet" method. This will automatically tell you if you have work to do or not and where to do it.

SPEC SHEET METHOD

Gather up all the Spec sheets for your gear. You can find them in the manuals for each device.

Find out what the Maximum Output of your mixing board is using the Spec Sheet. I have the Allen & Heath Qu-32 console at the time of this writing and the maximum Output of this board is +22dBu. Next find the maximum input and output of the next device in the chain.

If it is Powered Speakers – awesome! This may be greatly simplified. ALL you have to do at this point is turn up the input sensitivity until you have reached a good listening level OR you have reached "clipping" on the amp. Done!

Let's go with multiple devices in the chain, for example, a house graphic eq, crossover network device or DSP (driverack type device) then power amps.

In my case it is Mixer to the DBX Driverack 260 (My system controller) to the power amps. According to the spec sheet the maximum input and output of this device is +22dBu (factory settings). **PERFECT!** It is an **EXACT** match for the Output of my mixer. The Gain structure for the mixer and system processor is done! That was easy. This means my mixer and driverack will "track" signals together and reach the clip points simultaneously.

This is what that looks like:

A&H Qu32 Mixer = +22dBu, Driverack = +22dBu for Input & Output.

We haven't talked about SPEAKER RATINGS & AMP RATINGS yet but it is a good idea to make sure your speakers can handle what the amp puts out. Just because you find the CLIP point on your amp doesn't mean the speakers you have can handle the wattage. Do some research. **If only passive speakers had clip lights on them right?** Would be nice....

A general guideline for matching amplifiers to speakers is: 1.5 – 2 times the rated RMS power of the speaker. So if you have a 600 Watt RMS speaker box – you can safely place a 900W (1.5 x 600) - 1200W (2x 600) amplifier on them. You will often see numbers like this: 600W RMS / 1200W Program / 2400W Peak. **IGNORE the PEAK number**... it only means your speaker might be able to handle a peak load of that amount for a few microseconds before bursting into flames. It is also a number given on a lot of cheaper powered speakers to make it seem like you are really getting something powerful... it is nothing more than marketing hype.

The RMS number is a continuous number. Play all day long 24/7. This shouldn't be a problem, however, you can't do it forever...

The **Program** number is like live sound... lots of dynamics. It too is safe as long as you set your gain structure properly.

If your speaker can handle the amp at full power – that is great, but still do not let the amp clip as that can more than double the wattage to the speaker in a burst form and toast the thing.

An important thing to know is that an amp just below its clip point can produce its rated maximum wattage but once it goes into clipping (Full ON Red Light), the wattage exponentially goes through the roof! Just in the nick of time to fry your speakers. An amp occasionally bouncing in the red may not harm anything but this is where you should back down a bit to be safe.

If the speaker cannot handle the full wattage of the amp, you will have to turn the amp sensitivity down even if the amp isn't clipping. Review the spec sheets of the speakers and amps and you will know.

You could measure the voltage coming out of the amp with a TRUE RMS Multimeter and adjust the amp sensitivity until it reads the proper voltage. At that point you must STOP. Mark the spot with tape. You will have to do some math conversions of Watts to Volts to know the correct amount.

The formula for Watts to Voltage is the square root ($\sqrt{}$) of Watts X Ohms.

For example: 600W X 8Ω = 4800. The square root of 4800 = 69.3 Volts

Let's say you have two of the speakers paired together. That would be 1200 Watts total X a 4Ω load. That is the square root of 4800 = 69.3 Volts. No different there...
So with the mixer at UNITY (to start with) and a TRUE RMS meter hooked

up to the power amp, you would increase the input sensitivity knob until the amp reached 69 Volts. That is where you would stop so you don't hurt the speaker. With live music, the voltage is always jumping around. It is dynamic, so you may not get a stable reading if you try this live.

Word of warning: The amp is kicking out **voltage,** so be careful messing around back there. You can get shocked. It may not be 110 volts but it will still suck if you get bit. Be careful!

This is why it is important to match speakers and amps correctly otherwise leaving an amp wide open can ruin speakers. The amp is fine but the speakers can't handle the wattage.

My first mixer was the Presonus Studiolive 16.4.2. What a great entry level digital mixer. According to the spec sheet, the Maximum Output of the Presonus was +24dBu. That is +2dBu more output than my driverack INPUT & OUTPUT will allow. So right off the bat you can see that I would have to be careful not to push the mixer within 2dBu of clipping because other devices in the chain would clip first. This is what it looks like:

SL16 = +24dBu, DR260 = +22dBu

If you have ever heard "digital clipping" before, you know how horrible it sounds. It is nasty sounding. It is embarrassing too during a gig. I have totally done this a few times. So going from the Presonus board into the next device kind of puts it in a "bottleneck". Imagine a chain of devices with different input & output levels… it can get confusing quick.

When you match everything up (hopefully before purchasing), the gain structure will automatically be done for you and the end result will be you turning up the mixer, amps or powered speakers until the sound is loud enough without any clipping.

If there is a mismatch in levels between devices you have two choices:

1. Simply mark on the main fader the point at which the next device in the chain clips. This means you cannot or should not go past this mark.

2. See if a XLR in-line pad is available for the amount of cut you need so that your mixer lines up with your next device. This will allow you to use the mixer's full dynamic range. These in-line pads get inserted into your main outs on the console and pad the signal before hitting the next device. Pads smaller than 10dB seem to be harder to find now days. Do a search for them at Radio Design Labs. www.rdlnet.com

On to the Gain Structure procedure:

DOWN AND DIRTY METHOD

Word of Caution: DISCONNECT speakers from power amps. On the mixer turn off all eq's, gates, compressors, limiters on Channels and Main Bus. Make sure you have no limiters set to "ON" inside of your speaker management device (DSP, Driverack) as well.

Also, do the same on your playing device if you have those options. Leave playing device at its maximum setting…. I.e. Volume all the way up.

The reason I say to go ahead and do this is because you may deal with musicians who use tracks on a computer, iPad, iPod or iPhone devices to play intros etc. Often as sound guys we need them to send us their maximum volume on the device. Otherwise we have to crank the gains up on the console and all that does is add noise (hiss). Eliminate variables and have all musicians give you the full amount unless of course you notice clipping from their device. In that case, they would need to reduce the volume / signal.

Keep this is mind for ALL tests you perform otherwise you will have sub-optimal results. We want pure unaltered signal from the input to the output of the mixer thru the crossover thru the power-amps on out to your speakers.

GAIN STRUCTURE PROCEDURE:

Disconnect speakers. No sense in blasting large volume here.

Run Pink noise through channel on mixing board. Set channel fader at UNITY and increase gain (trim) to read meter 0dBu or -18dBFS.

Next, turn up Master Fader until you see main meter clip.

My guess is that you were not able to get the board to clip.

Now increase channel fader from unity to about +5dB or you could increase channel gain +5dB. Now push Master fader on up again until you see the main meter clip.

Look at next device – do you see any clipping there when you make the master fader output clip? If the mixer's output (dBu) is close or identical to what the next device will accept. The next device should be right at clipping as well.

Now take the master fader and move it up into clipping and down (out of clipping) while watching next device inputs and outputs to see if they are "tracking" together. Hopefully they are.

If the next device CLIPS before the Mixer's main meter show clipping.... You will have to mark the mixer's main fader with tape or something to let you know you cannot go past that mark. Even if you cannot see clipping on the main meter. Make sense?

This would tell me your mixer has a higher output capability than what the next device will accept. We should have been able to predict this behavior from the spec sheets. A lot like a tall load on a semi trying to go under a shorter bridge. There is going to be "Clipping".

Once you get the mixer and DSP tracking together all you have to do is turn up the amp input sensitivity knob until you see clipping. Make sure the main fader is up to the clip point or the next device's clip point, and then raise the amp knob up until you see clipping. **Mark that spot.** It is possible that your amp didn't reach clipping with the knob all the way up and that is fine. However, can your speakers handle the full output?

To summarize: You found the point on the mixer where the main output meter shows clipping.

You then looked to the next device in the chain to see if it was clipping. If it isn't you are good to proceed to next device or onto the power amp. If it IS clipping then that device becomes your **limiting factor.** You will need to mark the main output fader where the next device clips.

Last step is increase amp sensitivity until you see clipping on the amp. Mark with tape if clipping occurs before the knob is maxed out.

You should be able to raise and lower the main fader to "clipping" to see that the devices are now tracking together.

Turn down amp or shut off. Hook speaker cables back up. Turn amp back on or up and play some music through the system to confirm everything went well.

That is it! Your system should be talking the same language now.

SETTING LIMITERS:

Go to your limiter that is in control of the amp it is assigned to. Set to the "ON" position. Make sure you start with the threshold all the way up (+20dB). Push amp into clipping (disconnect speakers again of course) Slowly reduce threshold until the CLIP lights go OFF.

On my Driverack I use an OVEREASY Limiter which is like a "Soft Knee" in compression…. It starts to work a little sooner before threshold and then once threshold is reached it clamps down (Hard Knee). At this point, the limiter will not let the amps clip. The overeasy setting sounds smoother than a hard limiter.

Repeat for other amp channels. Done!

CONTROVERSY OVER AMP INPUT SENSITIVITY KNOBS:

Amplifiers are fixed gain devices, turning down the amplifier input knob or attenuator does NOT change the amps ability to reach full output. It just takes more INPUT voltage (from upstream) to achieve full power of the amp.

Many amplifiers will clip with an input signal over +6dBu when the sensitivity knob is all the way up.

Many mixers can deliver +18dBu of output before clipping.

This translates into reduced headroom by +12dBu if you insist the amp input sensitivity knob **has to be** all the way up.

It is "OK" if the knob is NOT all the way up. Perform a good Gain Structure and you will be fine.

I feel that this argument over the tiny knob would be irrelevant if more time were put into matching components correctly. After all, that is what the knob is for, to help match up incorrectly matched components.

BALANCING & TUNING THE SYSTEM RESPONSE

This section is going be long. I just don't know how to keep it short. Up to this point we have covered many things. I know that some if not a lot of it may be hard to fully understand but once you understand how the system response of a PA should be set up & what it is **supposed to look like** – it will all come together. Trust Me!

This is where you really make the system KICK BUTT. We had to go through many things in order to get here and we will probably have to go back and repeat or fine tune many things afterwards as well.

Keep in mind if you want to be great at mixing, you have to be great at setting up the system's response. This also means being able to get the system on the people and make it even or uniform throughout the listening space. That way you can mix on a nice clean canvas.

Here we go:

Many of you already know that the AUDIO SPECTRUM is loaded with frequencies right? 20Hz to 20kHz is a range I see often.

To be more realistic with live sound systems & especially the average bar PA, the spectrum is more like 40,50,60Hz on up to 10-14kHz.

20Hz is ridiculously low. 30Hz in a bar PA is really low too & it most likely isn't happening at any kind of serious decibel level. Anyway, we can debate that more later BUT it just depends on the quality of speakers you have. I do not have a set of $5,000 subs (each).

Have you ever *really* thought about the audio spectrum? What does that *really* mean? Ok, I know the audio spectrum has frequencies & those frequencies come out of the speakers. Did you know that there are lots of frequencies between 20Hz & 20kHz? There are more frequencies in the audio spectrum than the ones chosen for us on a 31 band graphic EQ.

How about 20,000Hz – 20Hz = 19,980Hz. That is a lot of frequencies. Did you know that all of those frequencies throughout the spectrum have to be at a certain volume or decibel level in order to be heard?

We often throw around jargon like 100 dB, 6 dB, 3dB….etc. This should give us an indication that this relates to volume or loudness. If there is such a thing as 100 dB, then there has to be such a thing like 0 dB. We can go even further and say -6 dB or -12 dB. This can be relative volumes (compared to) or absolute volumes. For example, the band is running at 95dB, A Weighted Slow @ 25 feet out. Hopefully this makes sense so far. Now while you are imagining that the audio spectrum runs from 20Hz to 20kHz, you can also imagine that all of those frequencies must be at a certain decibel level, meaning that each frequency has a loudness to it.

This is known as **FREQUENCY RESPONSE.**

Every speaker / PA system has some sort of Frequency Response going on. You can't see it & may not even be able to visualize it at this point but you can hear it.

Knowing what to do with this **FREQUENCY RESPONSE** is the MOST critical part of obtaining a Kick Butt Bar PA. Actually, any PA for that matter.

Let's take a look at an example here to demonstrate.

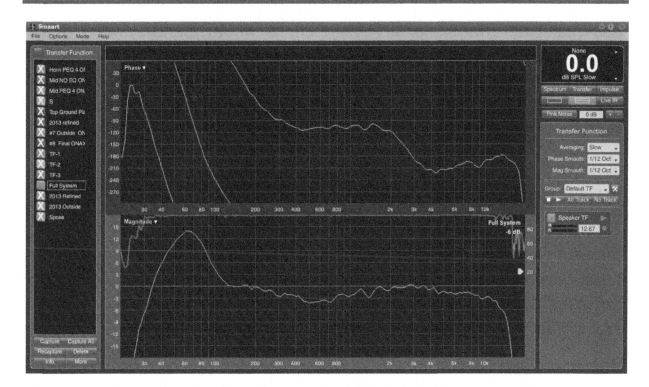

FREQUENCY RESPONSE (Amplitude vs. Frequency)

Here is a glimpse of a frequency response in real time. (Bottom window called "Magnitude"). You can see however, the large build-up in the low end and from around 100Hz on out to the high end of the spectrum – it is relatively flat. The area around 400-700Hz was reduced a bit more here due to the amount of instruments and vocals that will accumulate in this area. When this area was completely flat, it was just too much combined with the room, so I reduced it a bit.

It is common to have this low end build up due to the fact that it takes more volume in these frequencies to hear them the same way as the higher frequencies.

The upper window is the PHASE of the system. This system has pretty good phase coherence from about 300Hz to 2kHz. With the proper processing on this, the horn could have been in better phase with the rest of the system but with this DSP device, it was the best we could do.

The better the phase - the better the intelligibility of the system.

This real time analysis of a PA system is where the DILEMMA & the SOLUTION lies. The reason why a lot of sound systems DO NOT sound the way they are supposed to or the way you hope is because this audio spectrum is out of balance. In the Picture above, all the issues have been worked out using parametric Eq's and Sub amp level with time alignment.

Here we have in the signal path a mixer, graphic EQ, analog crossover, power amps & then a set of speakers. A lot can go wrong here with all the cabling and all the knobs. No wonder it is confusing.

Let me show you a case study that hopefully you will find enlightening. A friend brings me a PA and says "I can't get this thing to sound very good. The cabs don't sound the same. I bought these speakers used and was hoping they would do the job". I told him to bring the system over.

I listen to it and quickly agreed that one particular top box didn't sound the same as the other box. I decided to run through all of my checks: Examination and Verification of the Parts Plus Polarity….. sound familiar?

Once I used Dual FFT on it, it was obvious as to what the issue was.

Take a look:

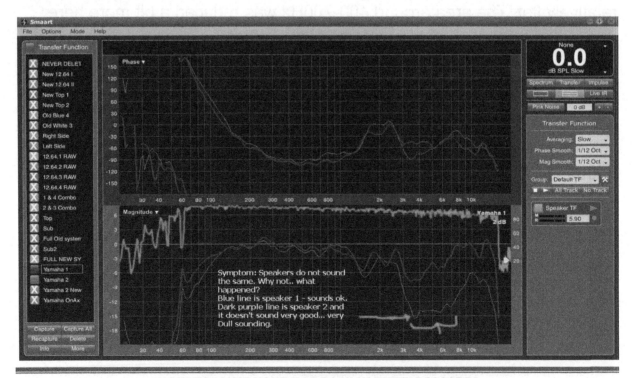

The dark purple line (speaker 2) has a horn that is not keeping up well with the blue line of the other box (speaker 1). A quick OHM reading on speaker two demonstrated the horn failing. Once the horn driver was replaced, the responses of the boxes match up better.

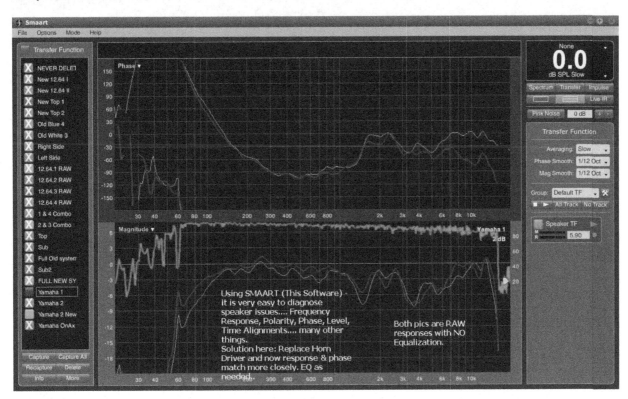

Replacing the horn driver helped to correct the basic function of the speaker but the response of the box is still RAW. With a little processing (parametric filters) on this, we were able to smooth out the response even more and make these Yamaha Club series speakers work well.

This would have been hard to diagnose and fully correct by ear alone.

Next example:

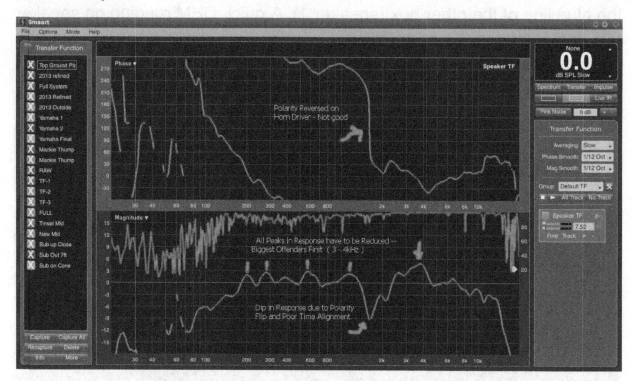

In this snap shot of a frequency response we have a Top Box that the owner says sounds different than the other box. Well, a quick look at the phase trace confirms immediately that this box has been messed with and the horn didn't get wired back up correctly. The phase has been inverted. Whenever you see a straight up and down vertical line on a phase trace (upper window), you know the phase is inverted. In a full range box, this could mean the crossover is failing from capacitor swelling. Possibly driven too hard at a show…etc…

I simply reversed the wiring on the horn driver and the phase trace shifted up by 180 degrees matching right up with the other top box. One problem solved! The other problems noted are those nasty peaks that need to be tamed with parametric eq's. I place parametrics right on the peaks and made cuts until the response smoothed out. Good luck trying that with a graphic eq.

The end result can be seen in the pic below. Now this is a response I feel is a decent place to start mixing on. Imagine trying to mix on a response of the above pic? Many do it all the time never giving it a thought of all the things that are wrong with the system to begin with. The mixer **cannot** solve those issues. Use the right tool for the right job.

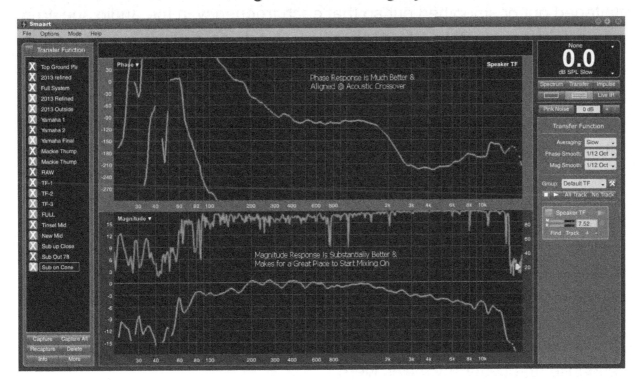

What I would like to stress to you is a fundamental "Starting Position" for setting up your audio spectrum. The industry standard strives for a SMOOTH PHASE / FREQUENCY RESPONSE. Look at any speaker "Spec" sheet and you will see the speaker's frequency response. Most likely processed with "their" processor.

On your system, imagine the frequency response "Smoothed" out if it already isn't. Get rid of those high peaks, set the balance between subs and tops, maybe even time align them. Once you mix on an optimized system, you won't want to mix any other way.

Historically, the word FLAT has been used & this seems to throw most sound engineers into a seizure. From a novice sound engineer's point of view, they think it means setting the graphic EQ & possibly even the channel strip EQ to 0 dB for all frequencies. I can see their concern, however that isn't what we mean here & they need to know that.

The true meaning from a SYSTEM TECH's point of view is that the actual ACOUSTIC frequency response coming out of the speakers has to be flattened out or smoothed out so that each frequency in the audio spectrum has a fair representation to each other. No hype or lack but with **equal energy.** It does get way more detailed than this but for now, the goal is equal energy.

A SYSTEM TECH, who has many roles, is an individual who thrives at the intricate details of sound transmission. With all the gear involved in a sound system and fundamentals like signal flow, gain structure (signal to noise ratio), driver alignment, level setting, phase & smooth frequency response, system techs recognize these concepts as absolute essentials to getting a PA at an optimal starting position **BEFORE** a sound / mix engineer even steps up to a mixing console. This allows the mix engineer a "clean slate" or clean canvas to work with. It is a challenge and when accomplished, very rewarding. For years I have been trying to master this area because I knew my mixing potential relied on it. The better the system response, the better my mixes were.

In the above pics before correction, the mix engineer gets to start mixing with a polarity flipped horn on one side (cancellation out front), 200, 280, 580Hz, 1.2kHz & 3.5kHz already boosted 3-5 dB & he didn't even know it existed. Yeah he could tell something was wrong because he was fighting the system on the diagnostic side of the mixer trying to make the mixer fix these issues. This makes for a long night of diagnosing and tweaking trying to get the system to sound right. The system tech's job is to locate & resolve these issues beforehand plus make sure the sound is on the people and that it sounds the same everywhere. For now, you get to be the system tech & the engineer.

The above scenario was measured without the graphic EQ engaged. If we release the bypass switch & inject it into the system and set all the frequency sliders to 0 dB, the frequency response wouldn't budge an inch as long as the graphic is at UNITY. At 0 dB, the graphic isn't boosting or cutting anything. It is essentially bypassed.

This PA system won't sound bad because we set the graphic EQ FLAT, it will sound bad because the Frequency response of the system is way out of balance to start with. The graphic had nothing to do with it. Just look at how messed up the response is and this is straight out of the box! These wild differences in volume across the spectrum make for a very lopsided sounding PA system & it is hard to correct during a show.

POINT: Your goal from here on out will be to Flatten out OR Smooth out your frequency response.

I call this my **"GROUND ZERO"**. My goal on every PA system I get my hands on is to get it to Ground Zero. When achieved, I know I am at the best starting position. Too me, it is a complete waste of time to mix on anything but an optimized PA that isn't at ground zero to start with. Unfortunately, it isn't always possible. I often go out and see other bands play to get an idea of what is going on in the area. I am usually approached and asked if I will help out to "Dial In" someone's sound system BUT I hate to say it, it is way too late in the game to do it justice. Not during a gig.... I have tried & honestly I am not very good at it. It is just impractical to unravel hours of missed examination and verification & correction in 5 minutes.

There are too many variables that affect the system & the mixing board didn't cause these problems and it also can't fix them. The damage is already done & the issue is further upstream. This is why we went through so much Examination & Verification of the Parts PLUS Polarity earlier. It had to be done.

I have a feeling that you are now starting to get the big picture here. Let's go back to the first pic I posted of a FULL System Response.

You may ask, "Why is the low end built-up and tapered downward like that?" That seems like a valid question so I will try to answer it the best I can. In sound system optimization you frequently learn things the hard way. When I flattened my first system, I did just that, I flattened it as flat as I could. I then took it out to the gig and was shocked that there wasn't any low end. I simply couldn't leave it that way so I ended up boosting the low end spectrum by 6-8 dB just to get thru the gig. It sounded better to my ears.

Completely FLAT -0- dB all the way across the spectrum just didn't sound good. All the nice low end was missing. When I asked my sound mentor friend what happened he explained to me that it has a lot to do with the way we hear, which is logarithmically. Live music typically has lots of low end build-up because our ears are less sensitive to the lower frequencies and they need to be louder. In the upper frequencies like 3.5 – 4kHz – our ears are most sensitive. Look up on the internet **"Equal Loudness Contours"**. Audiologists know the sensitivity of our ears vary for different frequencies. In sound optimization the low end build-up is called "Pink Shift or Spectral Tilt". Some even call it a "Haystack".

In my opinion, I feel the strength of the low end build-up is a personal thing & genre specific. Different styles of music require that low end to be adjusted accordingly. Rap music would require a very strong low end buildup while some classic rock may require a much lighter low end.

I had the opportunity to talk to a well known system tech that has been in the sound industry for 25 years and he explained to me that it is a genre specific thing for him. He mentioned he tuned a sound system for Snoop Dog and his sound engineer preferred a very strong low end build up. This build-up went from +20 dB @ TRUE 30 Hz with a Smooth / Flat response tapering down to 20kHz. Wow – that is crazy. I have never been to a rap concert before but I can only imagine what it would sound like. Lots of THUMP!

My current configuration for my bar PA is close to +6dB @ 40Hz and +14 dB @ 70Hz with a nice round bump at 70Hz tapering down to 100Hz (0 dB) and from there on out, I flatten the areas that need flattened out. This requires listening to the PA as you go to make sure an EQ cut is needed. Not all peaks need flattened. Dips should be ignored. You are shooting for a general trend of smoothness. You won't be able to make it perfectly smooth nor should you.

There is NO wrong or right way in setting the spectrum up, as long as it is sounds good to your ears. You can set it as strong in the low end as you want or as light as you want. I actually have programmed into my speaker management device three different low end build ups. There is a +8 dB, +12 dB & a +14 dB. Depending on the type of music, it is just a matter of toggling through the preset to see what suits the style of the band best.

Ok, now that you have a visual of the audio spectrum & the dilemma that causes a lot of PA's to sound bad, we need to talk about how we fix this. You now know what the "Dilemma" is and the goal so let's move onto the next topic.

SPEAKERS

Speakers come in many sizes. They vary in price quite a bit and they all have a Frequency Response. Some better than others. A speaker all by itself will demonstrate a wild response. They **ALL** have flaws of some sort. At this time, technology hasn't been able to produce a speaker that has a perfectly flat / smooth **phase** response. A great deal of time and money has gone into this and I constantly read that the ultimate goal of a speaker in an enclosure, is to have a flat smooth phase response right out of the box. Some companies are getting closer but the consensus is, it will never happen. In fact, it hasn't even happened with the additional help of an external speaker management devices controlling the speakers. There is still variance.

To accomplish our goal of a smooth response with minimal variance, we are simply not going to be able to leave the speaker untreated. We are gonna have to help it get to Ground Zero. So here is where we need to go next:

PINK NOISE

Pink Noise is the industry standard for sound system optimization. It is a very common test signal and will be called our KNOWN test source. There are a few "known's" in live sound so we need to at least start off with a reliable source and work through from there. Take a look at everything around you. Isn't it apparent that many things were built with symmetry? To get consistent angles and measurements, a known source had to be used. Carpenters use tape measures, speed squares, T-squares, levels & many other tools because they are KNOWN SOURCES. They are very reliable if used correctly. We too will have to use a known source if we want to see where the dilemma lies and to fix it.

Pink Noise is a test signal that has all the frequencies reined into EQUAL ENERGY per octave band. This is our KNOWN source.

USING PINK NOISE FOR SETTING FREQUENCY RESPONSE

Next, we finally get onto measuring the system. It's about time! Everyone wants to skip the essentials we talked about and get right into measuring. As you can see now, if we were to do that, our results may not be very good.

Ok, so what do we measure the system's frequency response with?

A measurement mic will do the trick. Behringer has a very affordable measurement mic, DBX makes a couple of models for around $100, Audix makes a very popular one called the TR-40 for under $200. You honestly don't need anything higher quality than the TR-40. The difference in quality between all these mics vs. high dollar measurement mics is their ability to measure the high & low frequencies. Many system techs use cheap ($100) mics and set the high end & low end to taste. You will be fine even with the cheapest measurement mic. I tested my DBX mic against an Audix TR40 and they tracked right with each other except around 5kHz & then they got back on track with each other on out. Slight differences, except the price of one is twice of the other. Pick whatever you like.

What we have to do next is find a way to measure your PA to take a peek at the frequency response. Hopefully it won't look like some of the pics above. It could look a lot worse BUT you aren't going to know until you dig in to them.

There are a couple of ways we can do this. One way is far superior than the other (analyzing program I used in above pics) but for now we will go with the basic inexpensive route. Unfortunately for those of you who have a basic analog crossover this could pose a problem. The reason being is that the new digital crossovers have a far superior "Tool Set" that will allow you to use PARAMETRIC EQUALIZERS instead of graphic EQ's to tune your system. Having the ability to "TIME ALIGN your system is also of vital importance. Some analog crossovers may have some of these features but I haven't seen any with both parametric EQ's and delay. So if you are a diehard analog guy, I hope you have at least a graphic EQ, preferably some parametric EQ's and the ability to delay your outputs on your crossover to time align your drivers. If not, then all of this could be a tossup.

Time alignment & a smooth phase / frequency response is the name of the game. If your tops have an internal crossover in them, you shouldn't have to align the horn with the woofer. You won't be able to anyways. It should already be done by the manufacturer but you do want align the top box with your subs for sure. In order to take advantage of this, you are going to need a device that is referred to as a **"Speaker Management System or Speaker Controller Device".** DBX makes several models at very affordable prices. Peavey makes one as well. EV has them, EAW has them. They are all over. I personally own three DBX Driveracks. One is on my FOH (Front of House) system (DR260) and the other two are on my monitor system (DRPA & DRPX). They correct flaws in the speaker response.

Recall again from the pics above… The out of balance Spectrum is what the PA system is doing all by itself. The graphic EQ is bypassed. What most sound guys do at this point is place a graphic EQ into the chain and start tweaking to get it to sound better. This can certainly help BUT we need to think past this mentality.

What we need to do is activate PARAMETRIC EQ's to handle the PEAKS in the speaker, balancing out the frequency response. A PARAMETRIC

EQ is a EQ that allows YOU to select the target frequency, the bandwidth (Q) of that frequency & the gain (how much to cut or boost). A GRAPHIC EQ like a 31 band EQ has already selected the frequencies for you. The Bandwidth called the "Q" is a 1/3 Octave. Of course, you get to move the sliders (the gain) up or down depending on what you want.

The issue with graphic EQ's not being the tool of choice is that you are limited to your target frequency. If your system's response is boosted for example at 720Hz, what choices do you have on the graphic to reach it? All you get is 630 Hz & 800Hz. At a 1/3 octave AND having to use two bands to reach 720 Hz, you can see quickly that this isn't ideal. What if 630 Hz is where it is supposed to be BUT 800Hz is your only option? You will end up using 800 Hz to reach 720 Hz and there is no doubt that if the cut is enough it will affect 630 Hz & even 1K. This could turn into a train wreck in a hurry! If a graphic is all you have, then you will have to do your best.

The Parametric EQ is a way different story. You can literally select 720Hz (within a Hertz or two, set how wide or narrow you want (Q), then apply the cut. It is like brain surgery. Very precise. It is the difference between a sniper rifle & a sawed off shot gun.

SEVERAL METHODS TO BALANCE SYSTEM

There are a few ways to balance the system. The first and simplest way is what we already covered in **"WHEN TIME IS UP – 3 BEST STEPS"**.

We took some headphones as our **known** source plus some good recorded music, and made the system match what we heard in the headphones. This takes practice and a good ear but the other methods are more technical when it comes to time alignment and being able to visualize what is going on. It is totally possible that using the headphones method failed and you just couldn't get the system to match up to the recorded music. This is where you move onto the next two methods. That is why I mentioned in an earlier chapter that if you have to "Cut to the Chase" and do not have time to go through everything properly, then try the headphone method and see if it helps. Let's say it didn't... Don't be discouraged... try the next two methods here:

The next approach will involve using "Auto EQ" (wizard) on driveracks or other speaker management devices. These devices were marketed as "Push the button" and when the system scans "the room", hit SAVE and you are on your way to being the best sounding band around! Well, it doesn't work that way. However, we can still use this method in a different way and achieve great results.

The last approach to Balancing & Tuning The System Response will involve a program like SMAART. It is also referred to as a "Dual FFT" analyzer. This type of measurement tool is what most speaker manufacturer's use in anechoic chambers to resolve speaker flaws and system issues. It is an incredible tool to optimize your sound system. However, you have to know how to use it and also know how to interpret the data.

Conclusion: You can skip over the next chapter on AUTO EQ WIZARD if you do not have a DSP, speaker management device or driverack. Some power amps now have DSP which will work great as well.

At a minimum, we may be able to tune the system using a graphic EQ plus a Dual FFT program like Smaart. It is not the best solution but it may be all you have.

TUNING SYSTEM USING THE AUTO EQ WIZARD

Take one stack of your speakers outside. That means, only one side. One Top and one Bottom (sub). Take a sub out and set your top on it just like you would at a gig. Point the speakers into an open space. We do this to eliminate reflections. You can only see a speaker's true character when there are no reflections. Reflections are copies of the original direct sound that come slamming back at the measurement mic (or our ear) and give us false readings.

The great outdoors give us a reflective free environment. Well, we still have the ground as a reflective surface but with measuring your top speakers 5 – 7 foot up in the air we will most likely get a decent original reading before any ground bounce comes into play. With subs, this could be a problem but it is a risk we have to take. We just have to do what we can since most of us do not have access to an anechoic chamber. An anechoic chamber is what the big companies use to achieve the flat frequency response we are looking for. It is controlled in every way: No wind, temperature is set just right, humidity & most importantly, no reflections.

We first used pink noise on a CD player or computer patched into the input channel of the mixer to set our gain structure. Right? Now we need to switch to using the pink noise generator on your speaker management device. There should be one built into it. Two separate pink noise sources that help you achieve two different types of things. Now, with a single stack of speakers outside away from reflective surfaces (as much as you can) take a mic stand with the measurement mic mounted on it and place it right in front of the speakers. Raise or lower the measurement mic until it is exactly in between the horn and the woofer of the TOP BOX. Most tops have a horn and a 12 or 15 inch woofer. Some have two 12's or even 15's in them. Maybe your tops are full range. For sake of simplicity (the concept remains the same), let's say you have a horn and a single woofer.

YOU WANT TO PLACE THE MEASUREMENT MIC EXACTLY IN THE MIDDLE OF THESE TWO. Place the mic there and you now have the correct height. So, how far out? That is a good question but it seems that most measuring experts think 8-10 feet is enough space for the drivers to combine & display even coverage to the mic.

Three to four times the longest dimension of the box will get you the proper distance. For example: the diagonal is the longest dimension on my TOP box and it is 29".

29" x 3 = 87" = 7'3"

29" x 4 = 116" = 9'8"

So you can place the mic somewhere between 7 & 9 feet out for this box. 10 feet out is good for most boxes.

If you are in too close or higher or lower than the other driver, one may dominate more than the other. If you are too far out (20-30ft.), you could get into ground bounce & the high frequencies of the horn taper down quick. This could cause you to mistakenly "Goose up" the high end more than needed & also make you want to EQ a dip that is really a cancellation.

Go 7 to 10 feet out on axis between horn and woofer (straight out) and set mic there. It is important that we focus on the TOP BOX for now and not the sub. The intelligibility of the PA will come from that top box. So we need to reserve the parametric Eq's for the biggest offenders.

Now we are ready to fire up the pink noise generator in your speaker management device. We are only going to measure one side because we will get a better reading. Once the optimization is complete, the work you have done will just transfer over to the other stack when it is time to hook it up. That is if your EQ's are linkable. You may have to re-enter your settings to the other side. Depends on the configuration of the unit.

When measuring: if you did this outdoors with both stacks, it would give you some crazy readings (comb filtering) that will cause you to make mistakes with the parametric eq's & it most likely won't sound good.

Break this procedure down to smaller chunks. One side only and one box at a time starting with the TOP box. Add sub only after you get Top box completed.

PINK NOISE FLATTENING SESSION OUTDOORS:

To do this method of optimization, you don't have to have your mixer hooked up. We are going from the crossover that has a pink noise generator built into it to the amps to the speakers. So our source is starting at the crossover instead of the mixer.

I am going to use a DBX Driverack PA for this test. I will try and explain the best I can. If you have a different brand of this type of device like the Behringer or Peavey speaker management device, try to employ the similar steps. Although slightly different, you should be able to pick up on the process.

Connect your measurement mic to the RTA INPUT on your speaker management device.

Push in the pink noise button by the RTA input to start the process. Hopefully your amps are up and marked where you were supposed to mark them during the gain structure procedure. Follow the onscreen directions. You may want to start going with the LOW precision setting first. It is quicker and it will be easier on you and the neighbors. Once you get the hang of it, you can move onto MEDIUM precision. I would get these down first before selecting the HIGH precision setting. This setting could take a long time to complete & it requires the system up close to gig level volume. The police could show up before you get it completed. Lol. If you live out in the country then you don't have to worry about this.

Alright, push in the little button that says RTA Input. Follow the onscreen directions. You will be asked to pick a response. In the Driverack, the responses are a "FLAT" setting or a "CURVED" setting. The curved responses are labeled, "A thru D". The curves are marketed as types of curves for a particular style of music such as a "Rock curve, or a Jazz Curve, Country curve"…etc. I would just go with the "FLAT RESPONSE" setting for now. Even if you have a different model, go with the Flat response setting. Why? A flat response will be less reactive in a reflective environment. Most club or bar acoustics are horrible to begin with. If you start using curves before you even know how your system reacts in a club, then this could be a problem. More on this later….

Now select LOW precision using the "wheel and button" mechanism on the driverack. Hit NEXT and start to raise the LEVEL on the unit. You will see on the screen what appears to be like a mini graphic EQ with a level setting. Keep turning up the level until you are at near gig level. It actually says "Set to performance Level". If you get into clipping, back down slightly and use the unclipped dB level every time for the rest of the session. Hit "next" & let it measure what your speakers are doing.

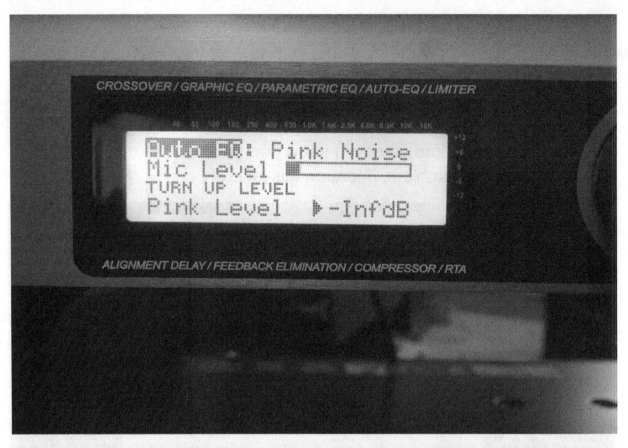

Once the scan is completed – it will show you the beginning screen again as if you never set to performance level. Not sure why but go ahead and "depress" the RTA Input button you pushed to do the scan. It will say "Auto Eq exiting".

Next, press "Program" to get out of the Wizard screen.

Hit "EQ". The results of the scan will appear on the EQ screen. What does it look like? Is it relatively smooth? Is it all up and down? Do you see any frequencies that were MAXED out?

Disregard any and ALL MAXED bands. The Wizard was trying to find those frequencies and couldn't so it boosted those bands…. Totally ignore them.

Focus on the most important part of the audio spectrum. That would be VOCALS. The range between 200Hz & 5kHz is a good start. This is NOT a definite range. Don't hold me to this. The reason I am giving this range is that you may not have enough parametric filters to EQ every flawed area. So you will have to prioritize. There is no need to place a parametric eq at

13kHz when that EQ filter could take care of an over bearing spike at 500Hz. Understand? So focus your attention to a smaller area of the spectrum (like vocal area) and work higher or lower only if you have parametric EQ's available. If you are able to use more parametric EQ's, then by all means go after another offender.

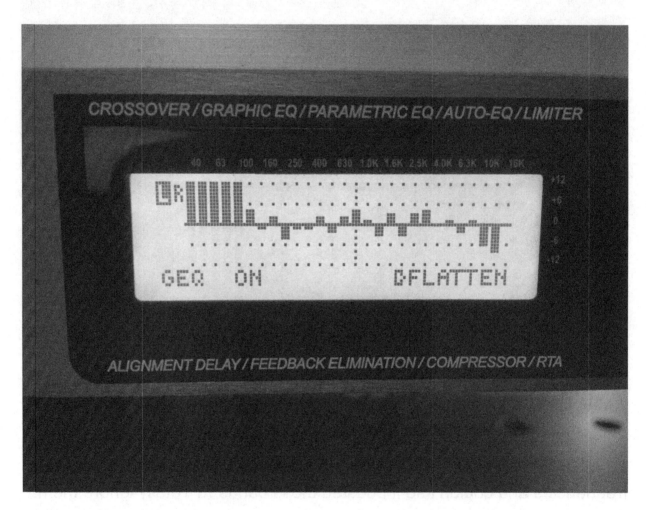

Did the procedure complete all the way or did a message appear on the screen that says, "NOT DONE". If it says not done, just ignore it and look at what the results are anyway. If more than FOUR EQ bands are boosted or cut in the max position, you will get a "NOT DONE" message. This simply means that the auto EQ couldn't find those frequencies so it kept boosting those bands in search of them. Don't worry about them. Most bar band PA's won't get down to 30Hz. 40Hz is really low too!

Don't get hung up on the lowest EQ frequencies. The same for the highest frequencies. Your PA most likely won't put out 16K or even 12.5K. Well, maybe 12.5K but not at any kind of real SPL. At this point, it is a non issue. Since you may be limited on parametric eq's – choose wisely and go after the biggest offenders first.

Using the RTA as a tool and the graphic EQ as a tool you can see what bands that are going to need to be adjusted (treated) with the PARAMETRIC EQ.

Do you see what we have just discovered? You have just discovered that your speakers are kicking out a certain frequency response with nothing engaged but pure clean signal. The response looks like you already have some kind of equalization on it. But you really don't have anything on it. This is the speaker's RAW frequency response. You say wait! Aren't my speakers already built and tuned to have the best response? Not necessarily. They could be but it doesn't mean that their response is going to be a flat frequency response requiring NO EQ whatsoever. That would be the ultimate speaker box. One that is measured and has a flat frequency response with NO EQ required, all frequencies reigned in with EQUAL energy right out of the factory. Just plug & play. How awesome would that be? Probably not happening though. There are some very high tech companies out there that still haven't achieved this and they still have to put a controller on the speakers to finish the job.

Have you discovered yet why randomly placing a graphic eq into the main signal path at the mix position may not be the best idea? Most people agree that a FLAT Graphic EQ is horrible sounding. Do you know why? Because it is! At least on the typical bar band PA it is. BUT, it isn't because it is flat. A flat EQ is really like saying there is NO EQ. The FLAT EQ is what your system is doing on its own whether that EQ was placed into the chain or not! Sound guys just inject it into the system and start tweaking the little sliders not ever knowing where ground zero is. You can't blame them for trying to make it sound better but this is the wrong approach. They ASSUME that since the graphic EQ says it is flat then the system must be flat and will sound lifeless so they start tweaking.

If you are fortunate enough to have new speakers that have the "Tunings" available for them then you may be able to skip all of this because the TUNINGS are really what we are trying to achieve by running the RTA unit outdoors on your speakers. The tunings are supposed to give you that flat frequency response that we desire. You may have to manually enter them into the device. Also, inside your speaker management device you may have in the data bank your speakers already programmed into it. Great, just select that. If you PA sounds good and you are satisfied.... Congratulations.

Even with all the possibilities for variation, the thing I like the most about my driverack is the ability to store your work. So set your system up the way the manufacturer says and save it. Listen to it. If it isn't what you expected, set up a different patch. You will then have to select "CUSTOM" for all of your choices. It's ok, just do it. If you are unsure where to set your crossover, just go back and read the chapter on "crossovers" and go with those settings.

Ok, so getting back to your first pass. After the auto Eq completes, take notice of what bands are boosted or cut. Any and all maxed bands boosted or cut, disregard. Go to the vocal range frequencies that are more pronounced & believable.

So in my case here, the five lowest frequencies need to be ignored. In fact, the sub wasn't ON so the auto EQ boosted those bands looking for those frequencies. These are not LEGIT. The higher frequencies I will ignore as well. Keep in mind, I only have a few parametrics to use so I will use them in the critical vocal range instead of up around 12-16kHz.

So what does all this mean? It means that your speakers are demonstrating some type of frequency response. If you plotted these 28 frequencies (for the driverack) out on a piece of paper and noted the boost or cut in dB, you could draw a line across those points and see what type of curves, boosts and cuts you have. This will give you an idea just how flat your frequency response is or just how jacked up the response is. Guess what? You didn't even need a graphic EQ to do that did you? It did it on its own and if you injected a graphic into this system now and set it to FLAT, it wouldn't sound any different. It would sound the same but the majority of sound guys would say it is because the EQ is set flat. Nope. Just unplug the EQ and say – "Ok, the EQ is now out of the chain so let's re-listen to it now…….wow it sounds exactly the same only the EQ is gone. I hope I have made my point clear on this GEQ subject now. It's not the Graphic EQ's fault……at least not for now.

So what do we use instead of the Graphic EQ? THE PARAMETRIC EQ (PEQ for short). Why? Because they are variable in every way. If you have a high peak at 3.57K you can set the parametric right at that position. You can't do that with a graphic EQ. With the graphic you get 3K & then 4K. You would have to use both of those frequencies to take out what is actually in the middle of those two. In the process, you will destroy frequencies that need to be left alone. This goes for dealing with feedback frequencies as well. You have to make wise decisions. The Parametric has the slick ability to set its bandwidth coverage called "Q". "Q" is how wide do you want the filter set to how narrow and specific do you want the filter set. You can't do that with the graphic EQ. Graphics are fixed and locked in at 1/3 octave which is a "Q" of 4.32. What are the chances that your problem area is going to be at 4K on the dot? Not very likely, so the Parametric EQ is what we use to correct the frequency response due to its powerful flexibility. The more parametrics you have, the easier it will be to achieve a flat response. Keep that in mind when purchasing a loudspeaker management device. Ok, so how do we use a parametric? Read on and you will see!

So here we are once again after your first RTA pass. So now what? Next, turn on your parametric EQ. In the DriveRack PA you have two parametrics on the subs & three for mids & highs (IF you are using FULLRANGE Top box). If your Tops are BI-AMPED, you can gain an additional two parametrics for your mids & still have three for the highs.

If it seems that I am jumping right into the unit, I am assuming you have read the manual and have a basic understanding of the parameters inside the device.

Before we start using parametrics, I want to go over an important concept of eq-ing that for some strange reason, took me a few years to grasp. I don't know why….

SUBTRACTIVE EQ-ING is how we should handle this tuning process. DIPS in a frequency response are much harder to hear than PEAKS. Therefore, we are only going to use a CUT or Subtractive EQ method. This means you will NOT boost ANY parametric Eq's. Only parametrics with CUTS should be used. After the system is tuned, feel free to boost an instrument on your **channel strip EQ**. Looking at the graphic display below, it appears 800Hz (see my marker line?) is boosted. In reality, the frequency response of this system is the INVERSE of this graphic eq display. 800Hz is actually CUT, NOT boosted. I hope this makes sense and I didn't just throw you way off. Again, the Auto EQ is BOOSTING frequencies in search of them. 800Hz was boosted on this first run because the auto eq was trying to find it when it was actually already cut. We do not need to concern ourselves with 800Hz on this display.

So for this method we are going to place parametric EQ's where the biggest cuts are at on the graphic display. I will name each frequency here that we should look at. Can you see the frequencies labeled at the top of the display window left to right? 40, 63, 100, 160…etc…

The three biggest offenders are:

200Hz, 1.25kHz & 2kHz. 500Hz is also an offender BUT we only have three parametrics. Knowing what I know about sound I would probably skip 2kHz and just know I may have a issue there later. Maybe I can handle that on the channel strip Eq? Maybe after I use the parametrics up, the graphic will be more flat and then I could use a filter there. We have some options.

The reason for possibly skipping over 2kHz is that so much of the instruments in the audio spectrum have 500Hz in them. Not to say that at 2kHz they don't but by the time 3-4 vocalists plus guitars add up, 500Hz is really built up. So I will go after 500Hz here so the system is less "Boxy" sounding.

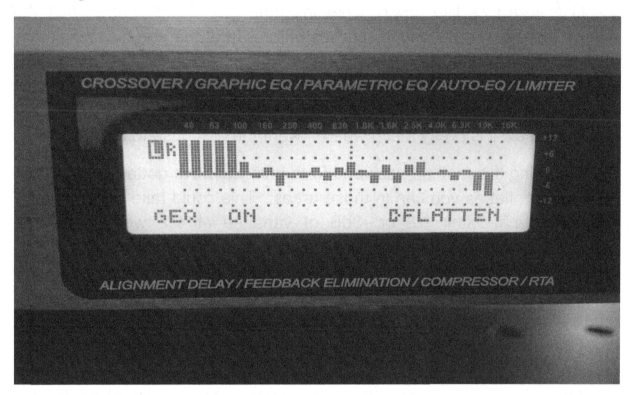

So I would choose to set my three parametrics right at 200Hz, 500Hz & 1.25kHz with a "BELL" curve because it is smooth and a bell shape. Select your "Q". "Q" is a difficult concept to wrap your brain around since you can't really see it in action on a screen. At least with this poor man's version you won't be able to see it. From here, it is trial and error. Go for a middle ground. The "Q" ranges from .20 (very wide bandwidth) to 16 (very specific like a pin point). So for the heck of it choose something like 3.41, 4.41, 5.71, or even 7.38 (respectively in DRPA) as your "Q". **Remember: a 1/3 octave graphic eq has a "Q" of 4.32** just to give you more perspective on "Q".

CUT the parametric in the SAME direction as the graphic or cut in the same direction as the graphic demonstrates. Here I cut 200Hz (-3dB), 500Hz (-

3dB) & 1.25kHz (-2dB). After selectively placing your parametric in the desired spot, zero out the graphic EQ (Not the Parametric) from your first pass and then while **leaving your newly placed parametrics as is**, re-run the auto EQ with pink noise. When that completes you should see a NEW CHANGE in the frequency response. What does the 200Hz, 500Hz & 1.25kHz area look like now? I bet the auto graphic EQ shows that the energy there isn't as hyped (or cut) as before. I bet the graphic shows less of a cut there (which was really a boost in the response) The parametric cuts made the change. After looking at my first pass, I added -1dB more cut to 200Hz and 500Hz. 1.25kHz was fine now. Next, I re-ran the auto EQ again and now it is pretty much where I want the response to be.

Keep working this method back and forth until you have gotten your system response as flat as you can in those areas. This could take some time. If you can get it within +3 or -3dB of variance all the way across the spectrum, you have done well. You have just achieved what is a highly desired response in sound system optimization. It does get more complicated than this but for now you are on the right path. So let's say you made your final RTA pass and when completed the auto EQ shows a nice smoother response across the audio spectrum. Congratulations! Your system WILL hopefully sound better and guess what? You used your graphic auto EQ in a way that you probably never thought of, a nice measurement tool. What you did is manipulated your system to use your graphic EQ as a tool to demonstrate to you what your speakers were saying (via live RTA) and then transferring that information onto your parametrics because that is what parametrics do best.

You could then add in the subs but I have found that the Auto Eq method using subs is difficult. Setting sub levels by ear is so much easier. Using your headphones with music will help you to set that balance to your tops.

Ok, hopefully you now have a system that is much more tame, more linear in its frequency response. Next step – SAVE IT! Theoretically speaking – you leave the parametrics where they are. Never change them UNLESS something changes in your system (got different speakers, more subs, more power...etc. OR maybe your response isn't as flat as you hoped and you just want to try to get it flatter) Hand write in a notebook ALL your settings so you will always have them in the event your unit crashes and you lose vital information. Hopefully that will never happen. Also, save the session in your DriveRack unit and name it. Call it "FLAT RESPONSE" or "DEFAULT" or whatever you like. It is your new starting position. All other Eq-ing you do can literally be done solely on the channel strip EQ's on your mixer. I have boat loads of graphic EQ's and I only use them in a worst case scenario if needed…. "Whisper singers & mic cuppers".

Hopefully, you can see now that we gave the graphic EQ a new starting point as well as your mixer. Here is my final pass for top box only:

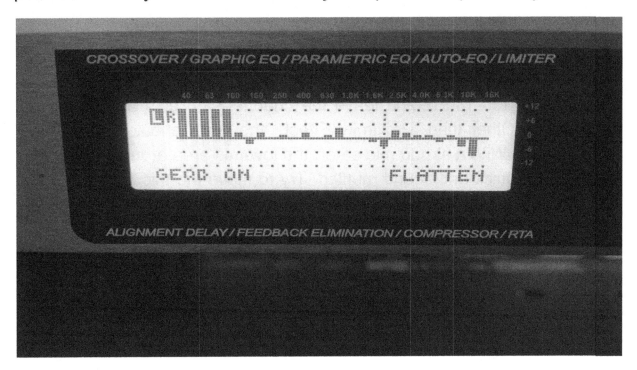

Now you could inject that graphic EQ into the main system set flat and people would say, "How did you get your system to sound so good when the EQ is flat?" Because the system was never flat to begin with! It was tagging along with the sound system's frequency response far from ground zero. Now it sounds good flat! Just think now, a 1.25k cut on the graphic by 3dB would really mean 1.25k got cut by 3dB whereas before 1.25k was already cut by 3dB (naturally) without the graphic, then you cut it again on the graphic by 3dB for a total cut of 6dB (natural). In reality, your graphic only shows a -3dB cut. The graphic display makes you think it is absolute BUT the natural response may not line up with it until you smooth out the response. I hope this makes sense...

What if 3k was naturally boosted +12dB because the natural frequency response was jacked way up there and you had to cut it completely on the graphic -12dB but was still unable to remove feedback from the vocals? Ok, you got it. Now that you are closer to an equal balanced energy system, your channel strip EQ's on your mixer have a new ground zero as well. A 80Hz +3dB boost will truly be a 80Hz +3dB boost. Thanks to our known source called Pink Noise we were able to make the response more balanced in energy than before.

Now run some music through this system. What do you think? Is your system much cleaner, tighter, more responsive? I hope so. Next, fire up a vocal mic and speak or sing into it. A good guide is to set a vocal high pass around 150-180Hz and leave the channel strip eq's at 0dB for any frequencies. Other than "Proximity Effect", the vocal should sound about right or close. With a strong mic eater, the low-mid area around 160-300Hz can build up quick and sound muffled. Try to move the mic away a bit and see if that low end muffle goes down. Use your ear to try and balance this out.

At this point, it should be nice and clear with plenty of power. Hook up a guitar and try that out if you have one. Your next step will be to take the system out to a gig and test it out. Worst case scenario, if you don't like how it sounds, you can always toggle back to your previous preset.

Alright, you are now at the gig and you have your system set up at the venue, what next? Well, we have a lot to talk about here. 90% of your great sound was actually achieved outside of the venue! You did this at home in your backyard or at a friend's farm or even your own farm. As you recall, we did this outdoors for a very specific reason, to remove as many reflective surfaces as possible. Reflective surfaces make your system react in a terrible way. You see, when you are at a gig and problems arise like feedback, weird phase issues, high pitched horns with ear piercing guitar or vocals or incredible muffled sounds, you are now on the DIAGNOSTIC side of the mixer. Your whole night is dedicated to stomping out fires that erupt out of nowhere.

So there are two parts. Number one: Your sound system. Number two: Your sound system in a room. You can't tell if it is the room causing the problem or your system causing the problem. One thing is for sure: If your sound system is the problem, the room is gonna make it worse. Usually due to reflections & room modes. Is it possible to make your system less reactive to even the most reflective environment? You already know this.... Your dang right you can. It is called a FLAT FREQUENCY RESPONSE.

A system that could care less what is around it is a system with a FLAT FREQUENCY RESPONSE. What a coincidence. We get to kill two very big birds with one stone! This is awesome because the places we go to the bar owner could give a crap about acoustic absorption and diffusers. And they wonder why bands are so loud and everything seems so lost and feedback is running rampant and the poor sound guy doesn't know what to do except work his trusty graphic Eq he has on the house so he starts cutting and cutting. Next thing you know people are complaining because everything is muffled and they can't hear the singer so the sound guy turns up the singer louder and then he discovers that he needs to make more cuts on his graphic to remove more feedback and the next thing you know everything sounds like DOG MEAT! Now what just happened? Good news, if you set up your system for the flat frequency response, it will be less reactive even in the worst acoustic environments. Any changes to your system in the venue will truly be caused by what the room did to it. Not the other way around. It will no longer be a circular problem.

So far we have made MAJOR changes to your system and hopefully you are awakened by the concepts and revelations that are taking place. 90% if not more of your system optimization is actually done outside of the venue. When you take your system in the venue it WILL be less reactive to its surroundings. You won't be fighting the system, only what the **room did to the system**. Now you say wait a minute, I thought you just said my system will no longer care what the room thinks. Well, yes, but it doesn't mean you can point your speakers wherever you want. Also, how loud do you plan on playing? There are physical barriers to consider and some rooms have a breaking point that no amount of sound pressure level will be able to override. We already know most bar & club owners don't even consider acoustical treatment as a valid business investment. I wish they did because our job would be so much easier. Their clients would appreciate it as well.

I would ask myself, "Just who I am trying to please". Hopefully, your audience and the band. I would also hope the band would say, "Our audience and our sound guy". Look, you aren't going to please everyone in the club. Someone is going to complain. There is no sense in pointing your speakers to the back corner of the bar where some patron has been sitting on the poker machine since you arrived to set up. He doesn't care about the band. He just wants to play poker & hit the jackpot.

I know this is a lot to digest and some of you may be overwhelmed. When the light bulb finally went off in my head with this stuff, I was really excited but then I actually got mad. I couldn't believe that (at least in my area) NO ONE knew this stuff. If they did, you couldn't tell by the sound of their system.

All these years of playing with issues, gig after gig, it just all caught up with me. What is ironic about all of this, is the fact that I thought I was going to figure all this out at the FOH mixer AND at the gig. It is funny looking back now to see that the mixer was no different than the GEQ. **They both get inserted into a system that is either at a flat frequency response or not.** Most Bar band PA's are not anywhere close to being flat. Everything I have done to my system didn't even involve the mixer. It didn't even involve the venue. It was the diagnostic side & it was all done in my backyard and driveway! It was all done with a speaker management device like the DriveRack and a measurement mic. What an incredible tool set to have.

This Auto RTA Wizard method works well but it is the "long method" and only requires a speaker management device and a measurement mic. So for around $300 - $1,000 depending on the model you get, you can correct your systems frequency response. Nothing else you buy for your system will compare to these tools in this price range.

Before we conclude this section you may be wondering when we get to the part where you run the auto EQ inside the venue. Many of you that have purchased a speaker management device did so on the premise (and the slick marketing ploys) that all you have to do is run this device at the venue and it will EQ the room. Really? The big boys on ProSoundWeb say the only way you could ever EQ a room is with a bulldozer! In other words – knock the walls down. Now the room is equalized! Before I ever learned about the flat frequency response system, I actually "pinked" a room. The place was full of people and when this thing was fired up, everyone covered their ears. We let it run and when it finished, I saved it. **Doing this on a system that is not flat to begin with - especially inside a poor acoustic venue is terrible! And that is where the fallacy is….**

It didn't sound very good and feedback was rampant. I haven't used it since to "pink" a room. Here are the reasons I feel that trying to pink a room may be a waste of time. The number one reason is the fact that your system is probably not flat to begin with. So essentially you are a running the auto EQ on a NON-flat system to begin with, in a room that is acoustically compromised. This is going to turn out badly.

Next, it is nearly impossible to get into a bar, club or venue where you can run this auto EQ without anyone else being there. All it takes is one person to walk in front of the system and your results are skewed. You can't have anyone there making any kind of noise. Even if all the workers were at the bar the smallest amount of noise will mess with your results. The next biggest reason is that you have to have a room that is acoustically sound to begin with. Most bars & clubs are not. They are highly reflective. When you use the pink noise generator on a room the system has to have a flat response FIRST (which at this point, you should have done by now) before firing up the pink noise generator. So when the auto EQ completes and demonstrates the results, you can see what the **room** did to your system. Did you catch that? Not your system BUT what the ROOM did to your system. What the results reveal is that some acoustic treatment is going to be needed. Since the owners are not going to acoustically treat their rooms then it is best to go in with a system that has a flat response and use your ear to EQ out room modes using good recorded music or your headphones method from the earlier chapter.

A lot of engineers use the processor to correct driver / system issues and once that is done, don't ever touch the device again. From there, they may have a graphic Eq on the front of house so they can "Tune" the system in the room. It is usually some cuts due to room modes. Then the channel strip to correct the source / mic combination.

Alright, time for me to wrap up this discussion…

Use manufacturer's Crossover settings AND Tunings if possible. They are always best in my opinion.

Use headphones with music to set tonal balance inside venue.

If you do not have any of those available… grab a speaker management device and dig in.

Here is a checklist:

	OPTIMIZATION CHECKLIST	
	EQUIPMENT / PROCEDURE	DONE
1	Absolutely need a Speaker Management System i.e. DBX Driverack PX, PA+ or 260 Peavey has a model VSX 26 Crown XTi amps have the "tool set" needed built right into the amp. Remember, we need the tools otherwise you can't optimize your system.	
2	Measurement Mic (Behringer, DBX RTA-M, Audix TR40…..)	
3	Pink Noise CD http://www.nch.com.au/tonegen/index.html	
4	Necessary cables & adapters to hook everything up	
5	Basic Crossover settings programmed into device.	
6	Gain Structure set with Pink Noise	
7	One stack of speakers outside (No reflections) Sub, Mid, High	
8	Measurement Mic between horn & mid woofer (8-10ft. out)	
9	Engage RTA auto EQ – select FLAT response – Low or Mid precision – turn level up (do not clip) note dB level and use every time till session is done	

10	View results – Any bands fully maxed (boosted or cut) – ignore. Correct biggest offenders first using subtractive EQ methods. This means you should **not** be boosting any parametric EQ's.
11	Turn Parametric EQ to the "ON" position. Choose "Bell" for Subs, Mids & Highs. Place parametrics where the GEQ shows cuts. Parametric cuts go in the same direction as the GEQ. Match level for level (dB) for starters. Choose a "Q". A low Q is broad for big changes. A high "Q" is more specific and narrow. This will take practice.
12	Zero out GEQ or select "Flatten". Leave Parametric settings in place. Re-run Auto EQ. View results. Make further adjustments. Keep repeating the process until you have clearly "leveled" out the frequency Response. +3 to -3 in variance is the acceptable range. The more parametrics you have available – the easier it will be to flatten the response.
13	Save your work in the unit PLUS hand write everything down in a Notebook.
14	Take system out to gig. Keep sound off the walls and on the audience.

Good Luck!

DUAL FFT

This last section of the book is an area of sound system optimization that I learned after everything we have talked about so far. I felt up to this point that I made major improvements and felt pretty good about my sound system. There was still a lot I didn't understand so I kept pursuing this sound mystery.

I became friends with a guy on one of the sound forums I frequent and he was explaining to me that he worked for a major sound company that does big shows. I had read several threads where he mentioned that he was trained in Smaart. Smaart is a dual FFT program (Fast Fourier Transform) that allows you to measure and monitor what your system is doing on the electrical & acoustic level simultaneously. Part of your system is traveling at the speed of light (Electrical) while the other part is traveling at the speed of sound (Acoustical). When you are able to view these signals on a laptop or computer in REAL time, it becomes incredibly obvious what a frequency response looks like. The potential to optimize your system just got way better as well. For once, you can SEE in real time what your system response looks like and what possibly needs optimized.

Most pro sound engineers have one of these Dual FFT programs and won't leave home without it. I know a way that you too can discover the power of a dual FFT analysis program like Smaart. There are other programs out there that basically do the same thing with some differences. There is SIM by Meyer Sound, Systune by Renkus-Heinz & EV has a program they use. There are many out there. "Smaart" by Rational Acoustics just happened to be the program my friend had and was trained on.

SMAART stands for: Sound Measurement Acoustical Analysis Real-Time Tool. To use this program, all you need is a computer and a USB / fire wire interface that has at least two inputs. I use the Presonus Firestudio Mobile as my interface. It was kind of intimidating at first but since I had done many outdoor sessions utilizing the "Auto EQ" method on my Driverack, I knew the goal was a Flat Frequency Response. Only this time, it was going to be in "Real-Time" on a screen and I wouldn't have to make a bunch of passes. I would then be able to grab a parametric EQ and place it on a peak and apply a cut while watching the screen.

If you are interested in this method I will take you through it. I highly recommend you purchase the software in the future because if you like what you hear so far then you will really like what you hear when your system is optimized with this tool. Like any tool, it can be used incorrectly so pay attention here.

If you would like to do the 30 Day Demo of Smaart, then we need to get situated first. There are some criteria here. Number one: A USB / fire wire device with at least two inputs. Interfaces like Presonus Firestudio Mobile, Roland OctaPre, Focusrite Scarlett 2i2…etc. will work great for this. Some of you may have these type of interfaces already if you are into studio recording.

On these interfaces, your measurement mic will go into one input and the other input gets a signal from your mixing board. With this method, we are going to use your mixing board this time. So we either have to use a "Y" cable to split the signal coming out of the main outputs or if you have additional ¼ jack outputs, you could use them. You could even use one side of the main out (L) that goes onto your crossover and the main out (R) could go into INPUT 2 on your interface. Keep in mind, we only need to measure one side of the PA (one stack only) for now.

So the signal comes out of your board and gets split with one side going into the input of your USB / Fire wire device. This allows an "Electrical" signal to be compared to the "Acoustical" signal (Measurement mic).

There are some great videos on YouTube from Rational Acoustics explaining how to correctly set this up. Take a look and see how they do it.

Here are a few pics of how it can be done on my mixer:

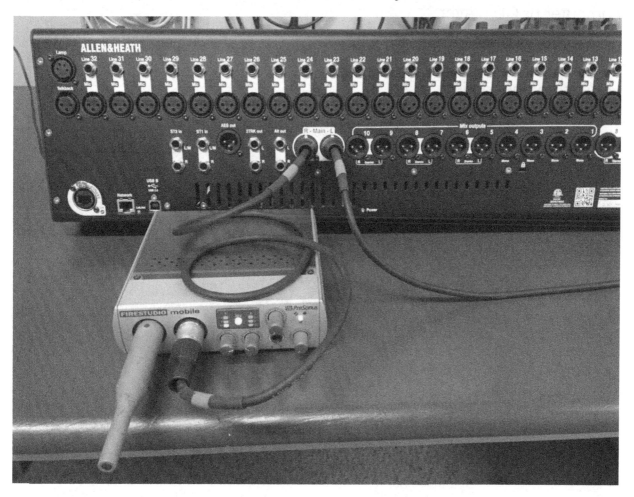

LEFT MAIN OUT goes onto Crossover / Speaker management device. RIGHT MAIN OUT goes to INPUT 2 on Interface. This is your "electrical" signal or "Reference" signal. INPUT 1 is where you want your measurement mic plugged into. Here I literally stuck it in input 1 just to demonstrate that ultimately, this is the input where the measurement mic goes. When you do this for real with a stack outside, you will connect a mic cable in between so you can run the measurement mic outside pointing at your speaker stack.

The Interface is then connected to your laptop or computer via Fire wire or USB, whichever you have. Pink Noise is generated within this mixer. Not all mixers have this capability. If your mixer does not have a pink noise generator, then feed pink noise into a channel on the mixer from your computer or CD player.

Here is another option:

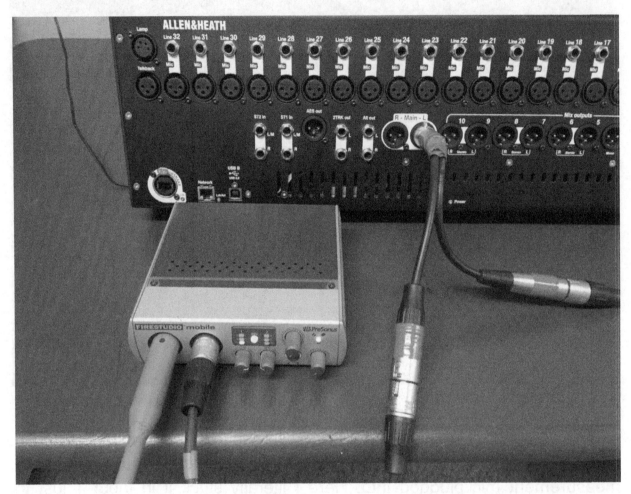

Using a "Y" cable has been common in the past. Very easy to use. The MAIN OUT is split. One side goes to Speaker Management device / Crossover and the other split goes into INPUT 2 on the interface. Measurement mic in Input 1.

I try to avoid using "Y" cables. There is confusion with them and how to use them properly. I wouldn't use a "Y splitter to run Smaart during a live show. I would use the "Alt. Out" or a "Matrix" on the console. There are better options.

If you are out in your backyard wanting to do some system optimization, by all means use a "Y" splitter if you have to. You are only going to use one stack of speakers anyways out of one side of the console. No harm there. The concern I have heard from others is the split side will suffer 3dB loss of signal compared to the un-split side. This may or may not be true. It is a mono split, not stereo, so keep this in mind if you use a "Y" splitter during a live show on a stereo PA system configuration. For now, we are not using smaart in a live gig situation. We are only trying to use it to optimize the system before the gig.

Here is another option:

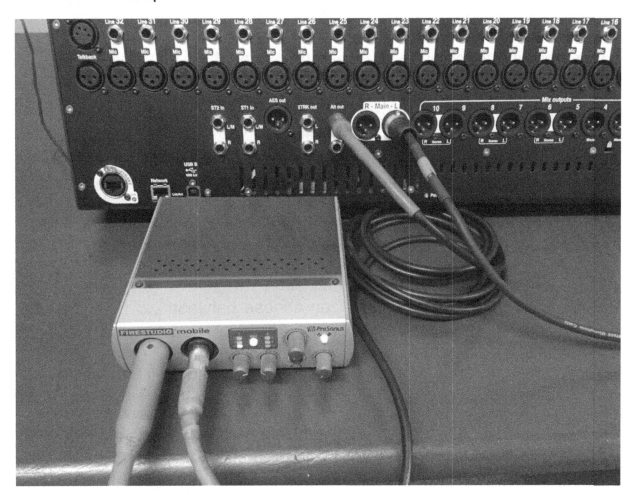

In this pic, I used the "ALT" out for my reference signal. The LEFT main out goes onto speaker management device / crossover like normal.

There are many ways to route this. This simple way is the 3 options I just listed here.

DOWNLOADING THE SMAART DEMO

Spend some time confirming you have everything you need before downloading the 30 day demo. **WARNING: Once you open up the program, the 30 day demo starts.**

Once the 30 days is up, the program trial period expires & LOCKS up and will not work unless you obtain a key which means you have to buy it for it to become functional again. The 30 day demo is enough time only if you have everything you need FIRST and the weather co-operates.

If you are still with me up to this point, then everything you have learned so far will soon triple. It is like when you are learning all kinds of MATH and you are told to do it a certain way but after you go through all the pain and suffering, the instructor says, "Now I will show you the easy way to do this". "Are you kidding! Why put us through all of that crap"! There is always a reason. So here is the Smaart method step by step. This is really slick!

Keep all of your other patches and start a brand new patch in your speaker management device. That way if you don't like how this turns out, you can always fall back to what worked before. Very important!

Download the Smaart demo: You can choose between v.8 (multichannel) or v.7Di (single channel). The single channel version is what I will use because it is simpler to demonstrate to a new user.

www.rationalacoustics.com

After you have your "Interface" connected to your computer and you know there is communication. Open the Smaart Demo. Make sure you have a speaker box hooked up and ready to go. I recommend just grabbing one of your tops for now without the sub. Set it up outside in the free field or if it is rainy and cold out, set it up in your garage or man cave…etc. Place the measurement mic just like we talked about in previous chapters.

Immediately upon opening Smaart v.7Di you will get this screen:

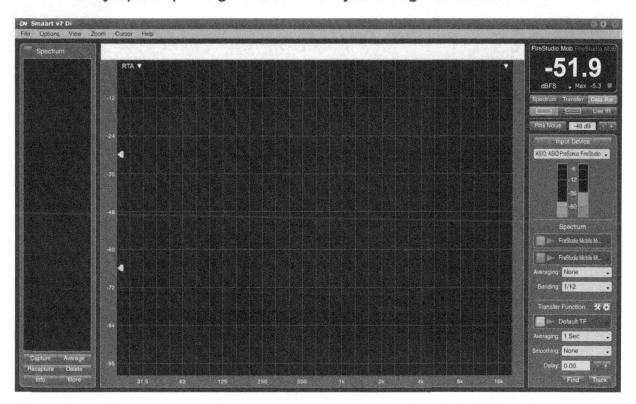

The YELLOW bar across the top indicates that this is a DEMO version. On a registered official copy the yellow bar is gone. One important thing to keep in mind is that the demo version will not allow you to save traces or "snapshots" of your frequency response. It also won't store the averaging, smoothing or delay settings on the transfer function. Each time you open the program, you will have to manually re-enter the settings. Not really that big of a deal, it is easy to do.

On this opening screen shot, it defaults to a single window "RTA" mode. Another nice thing that was automatically done is the "INPUT DEVICE" automatically found my PreSonus FireStudio Interface. Hopefully, your interface will be found automatically as well. If not, click on the little "hammer & wrench" icon just to the right of the "Transfer Function" tab. You could add "New Input Pair" or you can click the tab at the top of this window where it says "Default TF". Click on that tab and it will bring you to this window:

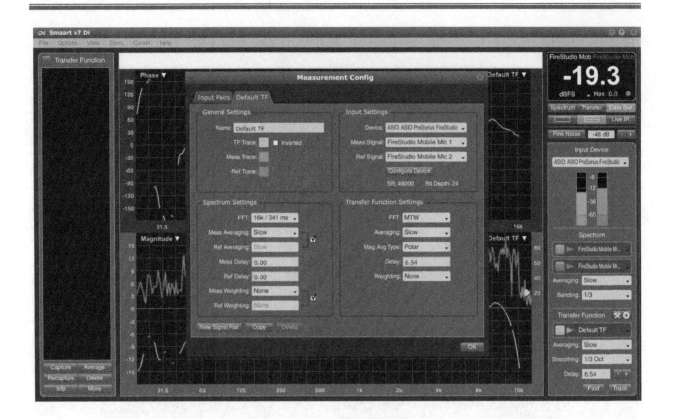

The top right shows current Input Settings. Click on the down arrow for "Device". You should see available devices. Since I connected a PreSonus device to my computer as my interface, I will choose that.

Some guys use their sound cards as an interface device. I have never done it that way before so I cannot really help you with that method. I do know that there are several ways to do this. Again, make sure you watch all of Rational Acoustics videos on YouTube on how to connect your devices. There is also a "manual" download for smaart on their website. The manual can help you with all the parameters and definitions of terms that you may have heard or haven't heard before.

Now, just below the "Device" tab is the "Meas Signal". This should be your INPUT 1 on your interface. It doesn't have to be but to keep it simple use Input 1. This is your Measurement Mic Input.

Just below Meas signal is "Ref Signal". This should be Input 2 on your interface. This is the line where you go out of your Main Out, Alt. Out or Matrix Out of the mixer and go straight to INPUT 2 on your interface.

If you accidently get these two inputs reversed, you will get crazy long "Delay Finder" times when you run the Delay Finder. I will show you more on this in a minute.

Ok, to measure the system using the Dual FFT mode we need to play pink noise through the PA. Go ahead and get that going into an input channel on the mixer. Place your MAIN fader up close to UNITY. You don't really have to go to UNITY but within -10. Un-mute the channel and bring up the fader until you can hear it coming out of the speaker. Fader position doesn't matter as much here like the MAIN fader position does. Remember, it doesn't have to be very loud. Next, go back to the smaart menu and let's select a few more things here.

Take your cursor and go up to the "TRANSFER" tab and click it. Next, go down to the bottom right where it says "TRANSFER FUNCTION".

Click on "Averaging". Change it to SLOW. Go to "SMOOTHING" and change it to 1/3 Oct. For now 1/3 octave will be better to use for a new beginner. It is smoother. As you get better using smaart, we can go to $1/6^{th}$ or even 1/12th octave resolution which will reveal in greater detail the flaws that exist in the frequency response. Let's get the basics down first.

After that is done, click on the "grayed" out arrow that looks like a "PLAY" button symbol on a CD player. It should turn GREEN indicating that it is ON. From here you want to increase your interface Input gains until the Green & Blue Signal levels in Smaart are just bouncing in the yellow on the meter.

You should now see an active trace climb up onto the screen showing you the frequency response of your speaker box. It may look chaotic at the moment but we will stabilize it with a few more moves.

The TOP display window or top half of the screen is your PHASE TRACE. What an incredibly important window and topic of discussion. Phase is EVERYTHING! SMOOTH linear phase that is…. More on this later.

The BOTTOM display window is your "MAGNITUDE TRACE" or you FREQUENCY RESPONSE of your sound system. Now you get to see in REAL TIME an active, breathing sound system response. Here is where we can allow the eyes to assist the ears in determining sound system issues and how to fix them. Very exciting!

A quick side note here: The crazy looking red line at the top of your magnitude trace window is your "coherence" trace. It is like a polygraph "lie detector" test. It tells you whether or not smaart is buying the data. If you see downward projections of red, it means it doesn't like the data in that area. It is false data cause by poor acoustics, wind and other noises. Even poor gain levels. If the red line is hovering up at the top and looks pretty smooth and settled, it is telling you that you can be certain that this data is reliable.

When adjusting signal levels on your mixer and interface unit pay attention to this coherence trace. Once the coherence trace gets close to the top, no additional input gain from your interface or volume from the PA is necessary. This means you can play pink noise in your backyard and tune the system without disturbing your neighbors. You can actually tune at low volume levels. This is not the case with the Auto EQ method. That requires near concert level.

Next, you want to click on the delay "FIND" button. Bottom right.

It calculates the delay in the system between your measurement mic and the reference signal (electrical) signal and compares the two. The Delay in the system should make sense. For example, my mic is about 7 feet out ONAXIS with the speaker. It would make sense that my delay time should be relatively close to 7 feet.

(Side note): In an earlier chapter we talked a bit about where to place the speakers & measurement mic. At the time of this writing, it is getting cold out again and it is also raining. I have no choice but to perform this exercise inside. I am using a shorter distance but 10 feet is always a good place to start.

After running the delay finder, it indicates the delay time to be 7.37ft. or 6.54ms. This makes perfect sense to me so I will click on "INSERT". It locks in the delay and suddenly my chaotic trace begins to settle into a slow "dancing" if you will, transfer function. If you run the delay finder and you get a delay time of 250 feet, ask yourself…. "Is my measurement mic about 250 feet out"? No, it isn't. That doesn't make any sense at all. Here is where you more than likely got your two INPUT signals reversed on accident. To this day, I occasionally get them swapped if I am not paying attention. Simply reverse the two cables to their proper inputs on the interface and rerun the delay finder. If these are correct and you still get a messed up reading, make sure your MAIN fader is up around -10 to unity otherwise the pink noise will not make it out of the mixer.

Moving on….. I hope you remembered to turn off ALL EQ's, gates, compressors, effects & high pass filters on the channel passing pink noise and also the master bus. After saving any previous work you have performed on your speaker management device. Start a "NEW" preset where you can make some changes and experiment with placing the parametric filters on the areas the transfer function shows possible flaws.

Rename your new patch "SMAART" or "NEW" or whatever you like. Here is where you get to experiment. This part of Smaart taught me so much about my sound system and the controls on my speaker processor… crossover, parametric filters, delay for drivers and even turning the input sensitivity knob up and down on a power amp and the affect it has on the transfer function.

Let the fun begin! Hopefully, you have a live Magnitude Trace happening on your sound system. Go ahead and grab a parametric Eq on your system controller and boost or cut it somewhere on frequency response. Pick 500 Hz for fun…. Can you see the change? Go ahead and sweep that same parametric down to 100Hz and then on up to 3kHz while watching the screen. Were you able to see the change? Hope so.

Now go to your crossover and play with the high pass filter on the top box. Top boxes are commonly high passed at 100HZ. Go ahead and lower the high pass down and watch the lower frequencies of the top box increase in magnitude (level). Now run the high pass up to 300Hz. Play with the SLOPES of the high pass filters. Go with an LR24 and then change it to a 18 and then a 12..etc… while watching the screen.

I am having you do these things so you can actually see for the first time the changes you are doing rather than blindly listening to what you have been doing. Again, we are using the eyes to help assist the ears.

Here is another one. Turn your amp input sensitivity up and down while watching the screen. What you should see is that the magnitude trace doesn't change in shape but it does change in LEVEL or AMPLITUDE.

This is important when it comes to balancing the Tops to the Subs. We have to get that balance correct before we align the two speaker boxes using electronic delay.

Very soon you will see why it is bad to adjust the crossover during a live show and the power amp input sensitivity levels. In a nutshell, it throws off the alignment and crossover point.

WHAT TO DO WITH THE RAW MAGNITUDE TRACE:

First and foremost: If you have a speaker management device with your speakers and the device is from the same company as your speakers, by all means – LOAD the presets up. Those are the "TUNINGS" the manufacturer found to be the best for the setup utilizing an anechoic chamber. It will be next to impossible for you to find better placed parametrics EQ's and level and delay than the manufacturer. So, go with manufacturer settings as much as you can. IF you do in fact have a complete system by the same manufacturer & have the factory presets loaded, there isn't anything initially for you to do here with smaart. You are optimized. You could continue to use smaart to see what using different EQ filters looks like or high pass filters. In other words, play around and learn.

What you could do however, is measure your system outdoors and take a picture using the "CAPTURE" function in smaart or use the "screen capture" function on your computer so you can save that snapshot. Remember, the smaart demo will not allow you to save traces so once you close the session, your snapshot will be gone when you reopen the program. If you use your computer's function to capture, you can then go back and look at it later. With that said, after capturing the trace, take the system into a room or venue and re-measure. You can then compare your free-field snapshot with NO walls or ceiling and compare it to the trace with your system in a room. You would then make changes (NOT in your system controller) but on a different EQ (main out) to bring the response back into formation of the free-field response. Once you take the system into a room, the room is going to change how the system sounds. It isn't because of the system. It is because of the room. Make sense?

You have two choices: Change the room acoustics using acoustic treatment methods. Most likely that isn't going to happen. Two: Make your system less reactive in that environment through use of EQ and level.

Now back to the discussion: If you have a speaker management device made from a different company than your speakers, you may have room to experiment here. If that device has some tunings in it that they say are for your speakers, go ahead and setup a preset with those. It won't hurt to give them a fair shake to see how they work for you. I personally find it hard to believe a speaker management device company (like DBX) would take the time to accumulate hundreds of speakers and find the best tunings. That would be very costly.

It is possible but unlikely that DBX or whoever could send out there processor to top speaker manufacturers in hopes those companies can come up with tunings to store in their units. That would cost a lot as well. Almost all top speaker companies make their own processing devices exactly for their speakers. It makes good business / marketing sense.

Sorry I keep digressing…..

Since we are here to **Make Your Bar PA Kick Butt!** You may not have all gear from a single manufacturer.

Here is a logical approach I have found after making tons of mistakes, that will help you get on the right track to optimizing your system. It is slightly different than my earlier comments about going for a flat or smooth response but you will see why in a moment.

How do you like the sound of your speakers with no EQ on them? Is it intelligible? Does good studio quality music sound intelligible through the speakers with NO EQ on them?

How does a singer's voice sound on your system with NO EQ on the main bus or processor? Also, let's use a high pass at 150-200Hz on the channel strip. How does it sound?

These are some good indicators on the performance of your system.

What we DO NOT want to do is start taking down peaks and filling valleys in your system response without listening to the RAW response first.

It may be possible that the RAW response of your system may sound decent once it is balanced with subs and time aligned. So before we get too far into applying parametric filters and graphic eq's, lets listen to you system with NO EQ on it whatsoever. If it turns out that there needs to be EQ applied, then we should add one filter at a time checking as we go to ensure we are on the right track. I would write down the EQ filters applied and what your reasoning is.... "I felt there was too much low mid when I say "TWO" on the microphone...etc... or my S's were way overkill through this PA... S's are up in the 8-10kHz range.

What you don't want to do is just start EQ-ing using smaart without listening first. I have seen so many "audio engineers" completely not care what the system sounds like until they have flattened out the response. Then they listen. I feel this is an error. The less EQ filters I use, the better off the phase of this system will be. Radical eq-ing messes with phase. However, proper eq-ing can help phase. So if you add a filter and it makes your magnitude trace look better but not your phase trace, I would think twice about using that filter. Use the coherence trace as a guide as well. If it doesn't like something or doesn't believe something, it will point to it with a red downward dip.

With that said, listen first & then decide what it is about the sound you don't like and try to correspond that area using smaart as a guide.

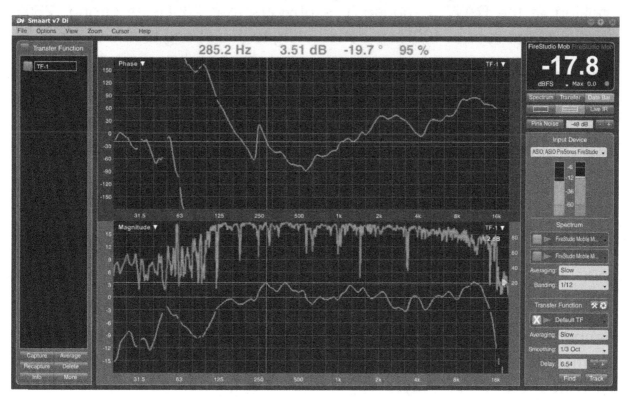

Take a look at a RAW system response of one of my top speaker boxes. It is not exactly smooth is it? The first reaction would be to take down some peaks. After you listen to this RAW response you find that good quality music actually sounds pretty good thru this system. Using a vocal mic doesn't sound bad either for NO EQ. However, there are a few areas that I think could be better.

A quick side note here: The crazy red line at the top is your "coherence" trace. There are a few areas of red pointing straight down on the areas it doesn't like. The data in this area is unreliable. Ignore making changes in those areas. The data can't be trusted.

While checking with a vocal mic, my system really stands out when you say "check one, **TWO!**" When I say **"TWO"** it is very strong and muffled in the low mid area. I am a close "proximity" speaker when it comes to mic technique. So "proximity effect" is playing a role here. Pulling away from the mic reveals that there is still a bit too much low mid in this system. Important observation here!

Looking at the RAW response it is obvious that 285Hz is strong. 285's neighbor next door at 400Hz is contributing to this strong muffle as well.

There is a dip in between 285Hz & 400Hz at 340Hz. I strongly recommend ignoring dips. Just like I mentioned in the "auto Eq" chapter…. We only want to go with subtractive eq-ing. A dip is so much harder to hear than a peak. Remember that!

As a second move after listening first, I will place a parametric "BELL" eq on 285Hz and make a cut to remove the peak.

I cut it -4dB with a "Q" of 5. Remember "Q" is how wide or narrow you want your cut to be. A "Q" of 10 would be really narrow. A "Q" of 1 would be really wide.

Rechecking the sound using the vocal mic, I found the cut at 285Hz seemed about right but still a little something in there. I have a choice of applying more cut OR maybe applying a cut at the next peak up at 400Hz. I ended up choosing a cut at 400Hz by -3dB and a "Q" of 6.4. Take a look at the screen shot:

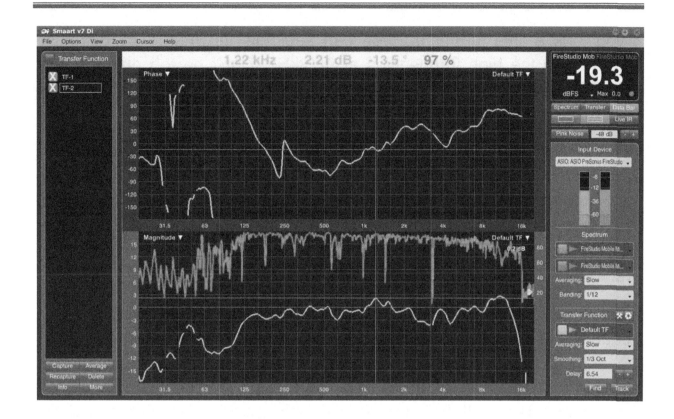

285Hz & 400Hz are smoothed out and after listening again with music and a speaking into a vocal mic – I liked what I was hearing better. It was more intelligible.

The next area that sparks my interest is at 1.22kHz. Smaart will tell you the frequency at the top (yellow bar) when hovering over the trace with your cursor.

What drew me to this area besides what the smaart screen shows is the "barkiness" in this area. Also, in a small reflective room when the system is brought up to gig level, I start to hear a "alienish" / reverb tone coming on. Eventually the system will start to feedback in this area if turned up really loud.

So I made a -3.5dB cut to the 1.22kHz area. If you look further up the frequency response, you can see a decent cut area at 2.45kHz – 3.25kHz. That is a natural cut in the response and I have NO desire to place a boost filter right there. It would more than likely make the sound worse.

If you look a little further up, you can see how the frequency response tapers up in the high frequencies at 9-12kHz. Don't be too alarmed with that. For people up close that could sound a bit sibilant but to those in the back row it may not sound bad at all.

For now I will not place a filter there and if I am at a gig and it is really sibilant on a vocal, I will want to make the cut on the individual channel strip but I may even make the cut on the main EQ. Most likely I will try both to see what works best for the mix. If I find myself always making cuts at specific frequencies at gigs, I think it would be a good idea to go to the system processor and take it out there.

Same thing goes for feedback frequencies. Keep a log book of feedback frequencies that you deal with at shows. If you find that when ringing out the main system or even monitors for that matter, that 2kHz is always taking off on you, go to the system controller and just take it out there. Save the setting. That will free up an EQ that could be better suited for something else.

If you have a log of 10 shows and the feedback frequencies are all different, you may not be able to make a specific cut in your processor. You may have to just deal with each room being different and requires ringing out every time.

Here is the final trace after applying three CUT parametric filters:

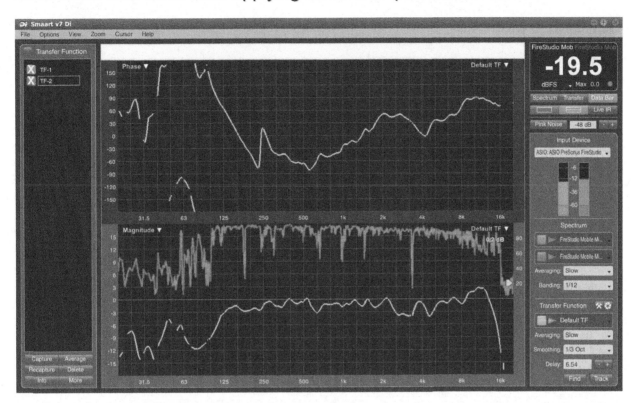

Looks smoother for sure.

Here is a BEFORE (RAW) response and an AFTER (processed) response:

BEFORE = Magenta

AFTER = Orange

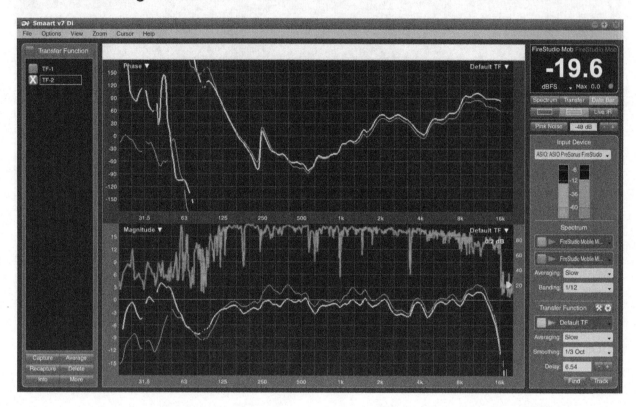

This is a good sounding TOP BOX right here! Don't get caught up in smoothing out a crooked line. You are going to have to ring out the system anyways and make cuts. What if the system goes into feedback in an area that is already dipped down naturally? You would have to apply some additional cut right there. Most likely a lot less of a cut since it is already dipped.

Let me try to explain in a different way... more things to consider before applying EQ filters:

Listen FIRST on a RAW un-Eq'd system. Take a snap shot of it with smaart. Ignore peaks and dips for a moment.

With vocal mics hooked up, bring system up to gig level and push system into feedback. Write down the first three feedback frequencies you get out of your system.

Go to your RAW smaart trace and see if there is any relationship to the peaks on the trace and the feedback frequencies you wrote down. Did your system feedback on a peak or did it happen in a dip?

If it happened on a peak, that may be a good indicator to make a cut there. If the feedback happened in a dipped area and you have to make a little more cut there, that is fine. So the natural dips in your raw response could possibly be a natural EQ cut for you without having to use a EQ filter (IF the feedback lands on it). Otherwise, you will end up placing EQ on top of EQ. That is bad!

Many guys flatten the whole response out and then ring the system out only to make a bunch of cuts right back into the system. **At least see if the feedback lands on a natural dip first.**

Well, there you have it for the TOP BOX. The high pass is set to 100Hz LR24 in the snapshot.

Next, we will add a subwoofer to this top box and balance it out and "Time Align" the TOP to the SUB.

ADDING THE SUBWOOFER TO THE SYSTEM:

Here is a pic of a subwoofer. The TOP box was muted so only the subs show.

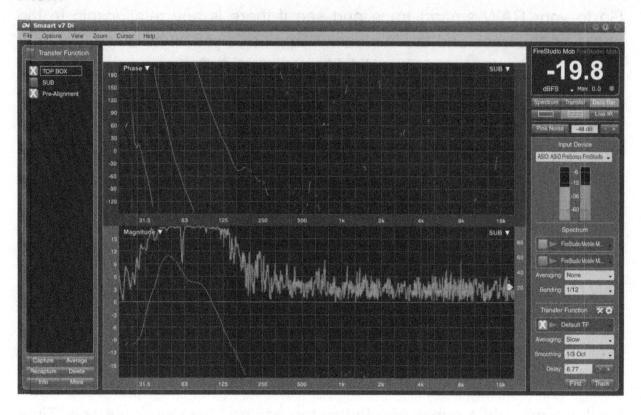

Here is a pic of a subwoofer added to the Top box...

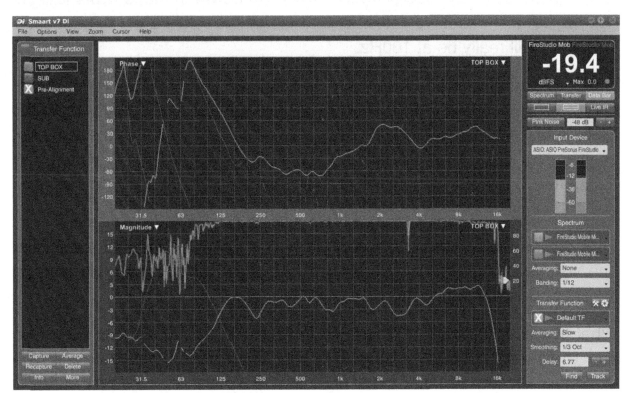

The Magenta line here is the TOP BOX. Notice the frequency response of the top box with a 105Hz high pass applied in the crossover.

The BLUE line is the frequency response of the subwoofer. It is hard to get good coherence in the subwoofer area since the low frequencies are omni-directional. Inside a room, the low frequencies build up everywhere.

Take a look at the upper window here. The PHASE trace window. Take notice that the magenta line (Top Box Phase) is to the right of the Blue (Sub Phase). This means the Top box is leading the sub in time.

To "Time Align" the top box to the sub, we must apply electronic delay. This delay is not like effects delay. It is electronic delay that allows us to slow a signal down.

Also in sound, we have electronic crossover points. For example, "my tops and subs are crossed over @ 100Hz". We hear this jargon all the time.

There is also such a thing as "Acoustic Crossover". This can be very different than electronic crossover. In order for us to accurately say my tops and subs are crossed at 100Hz, the level between those two speakers has to be the exact volume. At that point the electronic & acoustical crossover will really be at 100Hz.

In my case, and in your case for live sound, with rocks bands and such, we need the low end to be much stronger in volume than the top box. Probably by 6, 9 to 12dB louder than the top box. I talked about this in an earlier chapter. We don't hear low frequencies as easily as high frequencies. Our ears are logarithmic in hearing and it sounds more pleasing to our ears when the lower frequencies are stronger.

That is why you see the low end boost on the subwoofer trace. I only have one sub here but I would normally use two 18's per side. Usually about a 12dB boost compared to the Tops.

What this does is create a difference in what my electronic crossover says and what the acoustical crossover says. The stronger the low end, the acoustical crossover shifts to the right. If my electronic crossover is set to 105Hz and my low end is at +12dB, the acoustical crossover point is going to shift to the right and land somewhere in the 120-125Hz area. That is perfectly fine. What you don't want is that crossover point to get up too high into the 150-160Hz area.

What this means overall is that you may have to add a GAP in the crossover points. If I use 105Hz for the Tops and 105Hz for the Subs, this means my acoustical crossover is going to climb up to150-160Hz. I don't want that.

So it is common to add a GAP in between the crossover points. My Tops are 105Hz and my sub low pass filter on the crossover is set to 80Hz or close to it. This gets me the acoustical crossover point of 125Hz.

Let's take a look at the "Aligned Boxes" and I will show you how to do the alignment.

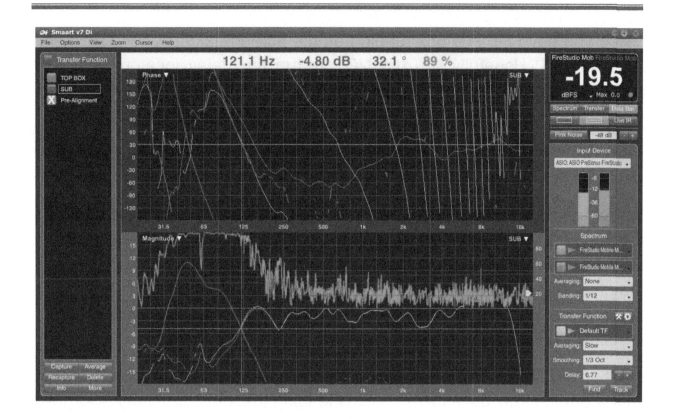

To align the boxes the first thing you have to do is measure the TOP box first. Once you engage the start button and adjust your gains on your interface you hit "FIND". This calculates the delay in the system and if it makes sense to you, hit INSERT. Your coherence trace should settle down and your phase trace should look smoother as well instead of lines running everywhere.

Once you lock the delay finder in place. **DO NOT RE-RUN THE DELAY FINDER! Also, do not reposition the measurement mic.** Very important…

Next, take a snapshot pic of your top box by hitting "CAPTURE" in smaart. This function is located in the bottom left of the program window. It will then store the trace in the column on the left.

Now mute the TOP box on your PA and fire up the Subwoofer. If you checkmark the little colored boxes in the smaart, that allows you to "hide" the trace from being viewed. Go ahead and hide the top box trace as well.

Once you have the sub box running on the screen, take a CAPTURE shot of it as well. Hopefully you did NOT re-run the delay locator since you turned the sub on. The delay locator is only for the top box in this application.

Once you have the Sub trace captured, Click on the top box trace or uncheck the box so it shows up on the main screen again.

It should look similar to what I have posted. Hopefully your sub volume can be set a little stronger than the top box can. If the subs cannot go up any higher, this means your top boxes are stronger than your subs. The only way to get the proper balance here is to decrease the volume of the TOPS (or add more subs). This can be done simply by turning the amplifier input sensitivity down until the balance is achieved.

Next, to start the alignment process, look to see what phase trace is in front of the other. In my case the top box is leading the subs. So delay is applied to the top box until it matches the sub box phase trace. The top box phase line should lay right onto of the sub phase line.

To do this, turn the TOP box back on or un-mute it so it will display on the screen again. MUTE or turn OFF the sub amplifier so that only the top box is live. Un-check the box for the saved sub trace. This places the sub trace snapshot back onto the smaart screen. Do the same for the top box trace.

So what you should see is a STATIC pic shot of the sub, the top, plus a live trace of the top on the screen.

Place your cursor at the EXACT crossover point on the two static traces. I have it indicated in the snapshot here. It is at 125Hz to be exact.

Leave your cursor there so that the "marker line" stays in place. Go to your output delay section on your crossover and start to apply delay in FINE millisecond (ms) increments while looking at your phase trace (upper window). Do you see the top box slowing down or moving to the left towards the sub phase trace? Hopefully you do.

What you want to do is make that top box phase trace sit right on top of the phase trace where you see your marker line located. See the crosshairs here?

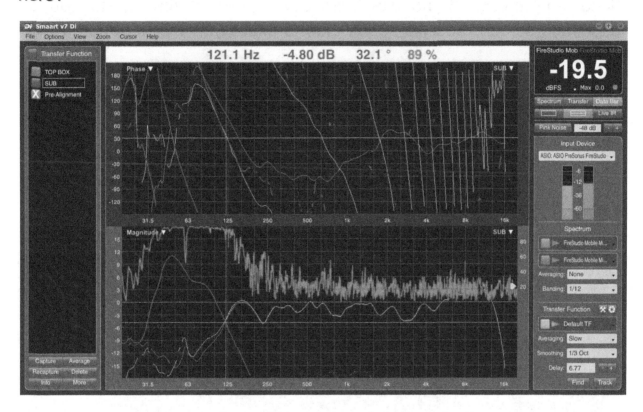

So the magenta line is the static top box phase line. The blue is the sub phase line and the orange is your LIVE in real time line that was on top of the magenta but is now on top of the blue phase line since you added delay. Man, I hope this makes sense? Very hard to demonstrate through writing…. It will start to make sense though… keep at it.

When you apply delay you will see your live phase trace start to fall apart. That is why you see all kinds of chaotic lines occurring in the live phase trace. Don't worry about that. Keep applying delay until your closest line to the sub trace lines up.

Let's reverse the situation: "My subs are in front of my TOPS". Ok… no problem. We have to add delay to your sub outputs.

Leave Subs live and running through smaart. Mute or shut off top box. Uncheck static traces of both top and sub so they are out on the screen.

So you should have a static capture of your top box and sub box on the screen PLUS a LIVE sub trace.

Place your cursor at the exact crossover point of the top to sub on magnitude screen. See pic again. Line is exactly on the target crossover point.

Apply delay to the SUBS output (NOT the top) and while viewing the screen keep applying delay until the sub trace lands on top of the top box trace. Especially at the crossover point.

You will not get it to be phased aligned everywhere. Just make sure it is phased aligned at the crossover point.

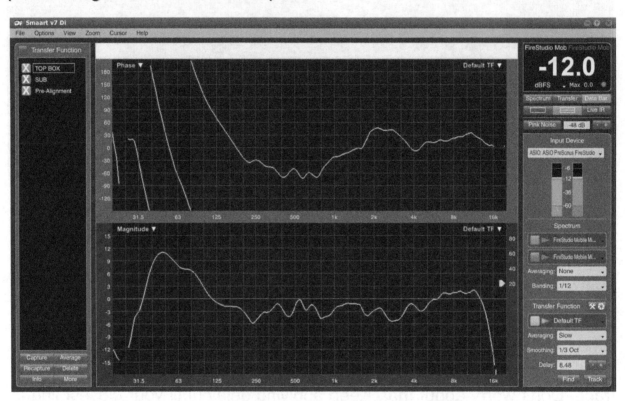

Here is the final trace. I did re-run the delay finder just so I could show you the final magnitude and phase trace of this system's transfer function.

It may not look as smooth as you would like but I am telling you to NOT get worked up over the dips you see here. Some of those natural dips are right where they need to be. Do not try to fill those dips. It would be better to reduce peaks of their neighbors if it was needed. Here I simply took a RAW response and reduced only the areas I felt needed addressed. That was done by playing music through the system FIRST and listening. It also involved using a vocal mic and utilizing "Proximity Effect" as a guide to set the low mid area of the response. Up very close, you would expect the vocal to be boomy. At 6 inches out, you may expect it to be a little thin. Go for a middle ground here.

A good guide for Top box is for a vocal to sound good with very little EQ on the channel strip and with a 150-180Hz high pass. Perhaps some 250-300Hz dip on the channel strip but other than that…. Basically flat Eq on a vocal. If you can achieve that – you should be at a good starting place.

Now it is time to take your system out to a gig and mix on it. Look out for room modes and handle those on your "Main Out" Eq. They are usually the low and low mid frequencies.

Don't forget that SMAART can also be used to optimize your monitor wedges. Most of the time however, Monitors simply need tonally balanced, Eq'd and rung out good while still maintaining clarity. That takes practice. SMAART can help assist if you are all out of ideas on how to make them sound better.

I think that will be it for the SMAART Demo. I highly encourage you all to purchase the software and take the class. Amazing stuff!

FINAL THOUGHTS

I hope this book was helpful in "MAKING YOUR BAR PA KICK BUTT".

There is a great deal of information to learn in this crazy field if you haven't already learned by now. We also haven't even talked about mixing that much either. What we did here was get your system to a place where you could mix better. A clean canvas! Basically, removing system issues and knowing when you are in a venue and problems arise that it is NOT a system issue rather it is a "Room Issue" or a "Musician Issue". It is one less thing to worry about.

This puts you on the creative side of mixing rather than the dreaded diagnostic side of the mixer. The result? I haven't messed with my Driverack in years. I haven't touched my graphic eq's in years. Of course with the new boards out now, you have full parametric eq's at your disposal.

This allows for some great shows! Keep honing your craft! Practice "critical listening" of your favorite bands. Things like tonal balance, panning, types of effects…etc… and try to mimic that. A kick drum should not sound like a cardboard box. A vocal shouldn't sound harsh or muffled… listen critically and don't be afraid to twist those knobs!

Be on the lookout for a DVD series on this book as well and hopefully it will help clear up any confusion that I may have created by not explaining things very well in this book. Sound is a hard thing to teach through writing, but I love it!

I will do a book or DVD on mixing as well in the near future.

I have to thank my wife Sherri & my children, Emily & Layne, for putting up with me always being gone doing shows or out in the garage studying sound and doing experiments with the PA. Also, for constantly talking about sound, PA's and gear. They have to be sick of it by now. It is a one of a kind lifestyle for sure.

I would like to thank my "sound" mentor Gary Perrett aka "Gadget" for taking me under his wing and being my friend all the way through this whole sound journey. We have met a lot of resistance out there in the sound field. Sound seems to be a very personal thing… Lol

Also, I would like to thank my father in-law Steve for all the support and help over the years. I could not have done any of this if it wasn't for those involved in this with me. I am blessed with good family and friends.

I cannot forget all of my band mates & musician friends that allowed me to try new things and experiment with their instruments and gear.

Big thanks to Andrew Vickers, Chad Green & Scotty Lawless for proof reading through this book and giving me advice. I really appreciate that!

Thank you to those who purchased this book, although possibly controversial in parts or doesn't make any sense. Keep trying… sound always depends….

Keep Rocking!

Dr. J

PS. If you have any questions feel free to email me at: jdepew309@yahoo.com

Glossary / Index

Anechoic Chamber – A specific room designed to have the least amount of reflections possible through absorption. It allows manufacturers to accurately measure and observe the behavior of sound waves. Ex. Measuring a speaker box. (pg.38,67)

Audio Spectrum – Basically the frequency range between 20Hz and 20kHz. (pg.53)

Bandwidth (Q) – How wide or narrow you want the space between two frequencies to be. A broad cut or boost would be a wide (Q) & a narrow specific cut or boost would be a narrow (Q). (pgs. 66)

Beaming – In sound, "beaming" can occur when it leaves the speaker. It is where the frequencies get real narrow and pierce like a laser beam. Some people describe this as an "Ice Pick" harshness on the ear drums. This usually occurs when the crossover of a speaker system is set incorrectly or not set at all. (pgs.39,40)

Boundary – The walls or structures in a building that can have an effect on sound waves.

Modes - Room Modes occur when you have a collection of resonances in a room that is excited by a sound system for example. These resonances "hang on" and overlap or mask the other notes being played. (pgs.12,14)

Crossover / Driverack / DSP (Digital Signal Processor) - Also known as a spectral divider. It is an electronic device that divides the audio signal into different frequency bands. These frequency bands go to different speakers such as a horn, mid woofer or a subwoofer. The end result is acoustic combining as if separate speakers are one giant speaker.(pgs11,12, 36,41, 65)

Cardioid – A *cardioid* microphone has the most sensitivity at the front and is least sensitive at the back. This isolates it from unwanted ambient sound and gives much more resistance to feedback than omnidirectional microphones. This makes a *cardioid* microphone particularly suitable for loud stages. (pgs.16, 19)

Continuity – Uninterrupted connection. (pg.20)

Direct Field - the sound you hear coming out of the speaker BEFORE it hits any reflections. It directly hits your ears first. (pgs 11,12, 15)

DSP - Digital Signal Processor – *See Crossover*

Dual FFT – Dual Fast Fourier Transform – is an important measurement method in the science of audio and acoustics measurement. It converts a signal into individual spectral components and thereby provides frequency information about the signal. FFTs are used for fault analysis, quality control, and condition monitoring of machines or systems. Smaart is a Dual FFT program. (pgs.38, 42)

Equal Loudness Contour - An equal-loudness contour is a measure of sound pressure, over the frequency spectrum, for which a listener perceives a constant loudness when presented with pure steady tones. The unit of measurement for loudness levels is the phon, and is arrived at by reference to equal-loudness contour. –Wikipedia definition. We hear logarithmically and the equal loudness contour represents that well.(pg62)

Feedback – Feedback occurs when outputs of a system are routed back as inputs as part of a chain of cause-and-effect that forms a circuit or loop. The system can then be said to feed back into itself. –Wikipedia (pg.16)

Flat Response – A response that is flat or smooth and has equal energy of frequencies all the way across the audio spectrum. (pg.12,)

Frequency Response – Refers to the frequency response of a speaker and the relationship of energy in each frequency across the audio spectrum. (pg.14, 54, 64)

Gain Before Feedback – In live sound mixing, *gain before feedback* (GBF) is a practical measure of how much a microphone can be amplified in a sound reinforcement system *before* causing audio *feedback*. - Wikipedia (pg.16)

Gain Structure – is where you calibrate the GAINS of each device in the signal chain to operate in its optimal range (best signal to noise ratio) plus set up the system where the meters on your mixing board accurately tell you what the meters on the rest of your gear are doing. So as you approach clipping on the mixer, you know you are approaching clipping on the next device in the chain and so on all the way to the power amps. (pgs. 36,44-52)

Graphic EQ – A fixed frequency device that allows you to only make boosts or cuts at predetermined ISO centers.

Multimeter – a device that allows you to measure many things like ohm ratings and voltage. This device is also used to check continuity, for example, a complete circuit or a break in the circuit when checking microphone cables or speaker cables.(pg.23)

Parametric EQ – an eq that allows you to adjust all parameters. You can choose your frequency, how much you would like to boost or cut that frequency and how wide or narrow you would like that boost or cut to be. Parametric EQs are the standard (pg.24, 37)

PFL – Pre Fade Listen – Listening to the signal Pre Fader or before the fader. (pg.13,)

Pink Noise – A Known Test Signal calibrated with continuous random noise having equal energy per octave across the audio spectrum. (pgs. 45, 64)

Polarity - Polarity is either IN or it is OUT. It is ABSOLUTE. Zero degrees together or 180 degrees apart from each other. Positive polarity is the same as two race cars going around the track TOGETHER neck and neck. The Input is moving in the **same** direction as the output. (20,31,32)

Phase – Phase in sound systems refers to the relationship of separate speakers systems and their phase. They could be offset from each other causing a delay in one signal or the other. Often in sound system optimization, electronic delay is used to align these phase offsets (41,42)

Reference – A KNOWN source. Pink Noise is a great example. A Test Tone is a calibrated tone used for many applications.(12,)

RTA – Real Time Analyzer – a device that allows you to see the perceived loudness of frequencies in the sound spectrum or in your surroundings. It does not take "time" into consideration.(16,)

SMAART – Sound Measurement Acoustical Analysis Real-Time Tool – a software program that allows you to measure and analyze audio signals and compare their relationship to each other. One signal is traveling at the speed of sound (acoustic signal) and the other is traveling at the speed of light (electrical signal). This allows you to make adjustments to the timing of these signals as well as adjust the frequency response of these systems with EQ. Smaart is also used to analyze room acoustics and various other features. (89,90)

Speaker Management Device – *See Crossover*

Speakon NL2 / NL4 Connectors – (29,34,35) A type of connector made by Neutrik (mostly) that "locks" into place preventing the cable from coming out prematurely. They have the NL2 for two wires and the NL 4 for four wires. They also make an NL8 that allow eight wires to be ran simultaneously off of one connector. For example, a four way PA system.

Spectral Tilt / Haystack / Pink shift – Where the frequency response of a sound system is purposely tilted in favor of low end information. (pgs. 40,)

Test Tone – a specific calibrated tone used for testing or measuring in sound systems. Test tones are also used in hearing tests (41,42)

Tuning – Tuning is the process of adjusting the sound system frequency response through level, polarity, time & equalization. The goal is to remove flaws in the system so you can actually mix on a sound system properly. (11,12, 37,

Time Alignment – the process where the individual sound systems (Tops, Subs, Sidefill & lipfills) are joined together in phase. This is accomplished using Smaart or some other Dual FFT program. (37,41,42)

Unity – Unity defined in live sound is nothing more than 0dB between two electronic devices. The Inputs and Outputs are at a 1:1 ratio. One Volt in equals one Volt out. The Output level is the same as the Input level.

Unverified – Not verified. An assumption. Not in a verified known state. Ex. "I assume the speaker is wired correctly since I bought it off of a friend who does sound." Don't assume anything!

Made in the USA
Las Vegas, NV
04 April 2024

88266068R00070